Edexcel GCSE

Mathematics B Modular Higher

Student Book
Unit 2

Series Director: Keith Pledger
Series Editor: Graham Cumming

Authors:
Chris Baston
Julie Bolter
Gareth Cole
Gill Dyer
Michael Flowers
Karen Hughes
Peter Jolly
Joan Knott
Jean Linsky
Graham Newman
Rob Pepper
Joe Petran
Keith Pledger
Rob Summerson
Kevin Tanner
Brian Western

A PEARSON COMPANY

Published by Pearson Education Limited, a company incorporated in England and Wales, having its registered office at Edinburgh Gate, Harlow, Essex, CM20 2JE. Registered company number: 872828

Edexcel is a registered trademark of Edexcel Limited

Text © Chris Baston, Julie Bolter, Gareth Cole, Gill Dyer, Michael Flowers, Karen Hughes, Peter Jolly, Joan Knott, Jean Linsky, Graham Newman, Rob Pepper, Joe Petran, Keith Pledger, Rob Summerson, Kevin Tanner, Brian Western and Pearson Education Limited 2010

The rights of Chris Baston, Julie Bolter, Gareth Cole, Gill Dyer, Michael Flowers, Karen Hughes, Peter Jolly, Joan Knott, Jean Linsky, Graham Newman, Rob Pepper, Joe Petran, Keith Pledger, Rob Summerson, Kevin Tanner and Brian Western to be identified as the authors of this Work have been asserted by them in accordance with the Copyright, Designs and Patent Act, 1988.

First published 2010

13 12 11 10
10 9 8 7 6 5 4 3 2 1

British Library Cataloguing in Publication Data
A catalogue record for this book is available from the British Library

ISBN 978 1 84690 807 1

Typeset by Tech-Set Ltd, Gateshead
Picture research by Rebecca Sodergren
Printed in Great Britain at Scotprint, Haddington

Acknowledgements
The publisher would like to thank the following for their kind permission to reproduce their photographs:
(Key: b-bottom; c-centre; l-left; r-right; t-top)

Alamy Images: aerialarchives.com 92; dbimages 128; **Corbis:** Deborah Betz Collection 121; Visuals Unlimited 13; **Getty Images:** Adrian Dennis 25; Digital Vision 207r; **iStockphoto:** Alena Brozova 208c; Kirill Putchenko 206-207; Nicole S. Young 1; **Photolibrary.com:** B Harrington 177; Jutta Klee 140; **Reuters:** Luke MacGregor 40; **Rex Features:** Ray Roberts 153; Sipa Press 79; **Science Photo Library Ltd:** Dr L. Caro 47; Mark Garlick 66; Kevin A Horgan 59; **Shutterstock:** 208-209; Nicola Gavin 207l; Ivanov 209r; Marjan Veljanoski 209l.

All other images © Pearson Education.

Every effort has been made to trace the copyright holders and we apologise in advance for any unintentional omissions. We would be pleased to insert the appropriate acknowledgement in any subsequent edition of this publication.

Disclaimer

This material has been published on behalf of Edexcel and offers high-quality support for the delivery of Edexcel qualifications.
This does not mean that the material is essential to achieve any Edexcel qualification, nor does it mean that it is the only suitable material available to support any Edexcel qualification. Edexcel material will not be used verbatim in setting any Edexcel examination or assessment. Any resource lists produced by Edexcel shall include this and other appropriate resources.

Copies of official specifications for all Edexcel qualifications may be found on the Edexcel website: www.edexcel.com

Contents

About this book

All set to make the grade!

Edexcel GCSE Mathematics is specially written to help you get your best grade in the exams.
Remember this is a non-calculator unit.

Section objectives show what you'll be learning.

Recap with a skills check at the start of a section – make sure you're up to speed.

Crystal-clear worked examples – step-by-step guides to answering questions correctly, with helpful hints and reminders.

Loads of practice to help you feel secure before you move on.

Full coverage of the new-style assessment objective questions – AO2 and AO3.

Graded questions – so you know what you're achieving.

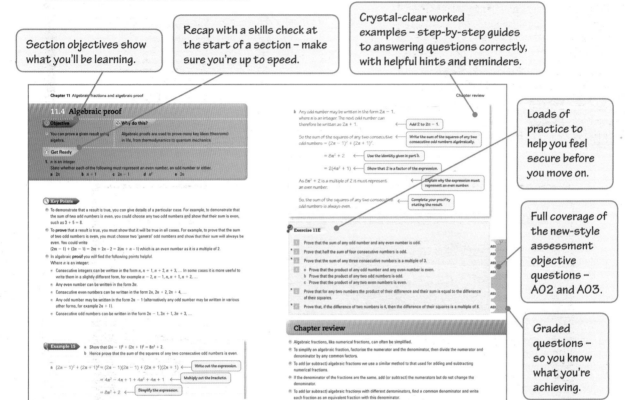

'Focus on AO2/3' pages demystify the new assessment objectives.

A fully worked example of an AO2/3 question... ...makes other AO2/3 questions on the same topic easy to tackle.

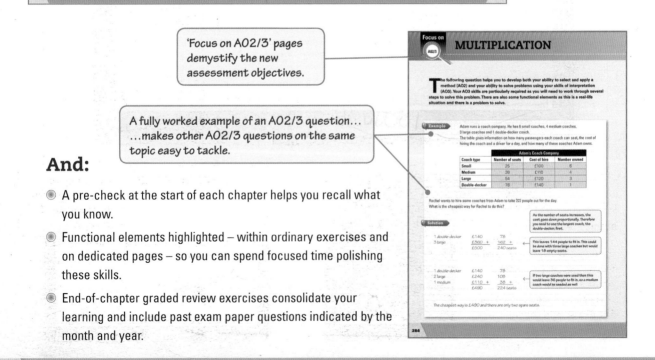

And:

- A pre-check at the start of each chapter helps you recall what you know.

- Functional elements highlighted – within ordinary exercises and on dedicated pages – so you can spend focused time polishing these skills.

- End-of-chapter graded review exercises consolidate your learning and include past exam paper questions indicated by the month and year.

About ActiveTeach

Use **ActiveTeach** to view and present the course on screen with exciting interactive content.

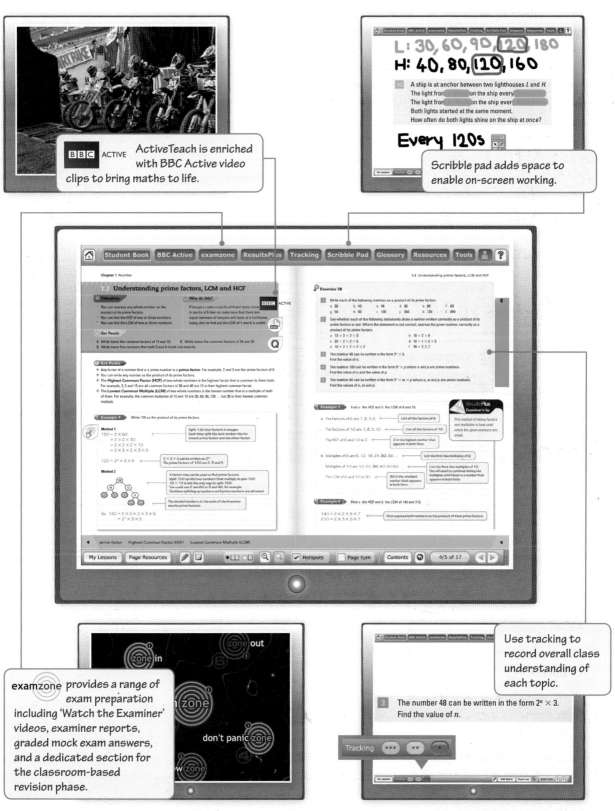

BBC ACTIVE — ActiveTeach is enriched with BBC Active video clips to bring maths to life.

Scribble pad adds space to enable on-screen working.

examzone provides a range of exam preparation including 'Watch the Examiner' videos, examiner reports, graded mock exam answers, and a dedicated section for the classroom-based revision phase.

Use tracking to record overall class understanding of each topic.

About Assessment Objectives

Assessment Objectives define the types of question that are set in the exam.

Assessment Objective	What it is	What this means	Range % of marks in the exam
AO1	**Recall** and use knowledge of the prescribed content.	Standard questions testing your knowledge of each topic.	45-55
AO2	**Select** and apply mathematical methods in a range of contexts.	Deciding what method you need to use to get to the correct solution to a contextualised problem.	25-35
AO3	**Interpret** and analyse problems and generate strategies to solve them.	Solving problems by deciding how and explaining why.	15-25

The proportion of marks available in the exam varies with each Assessment Objective. Don't miss out, make sure you know how to do AO2 and AO3 questions!

What does an AO2 question look like?

D AO2

This just needs you to
(a) read and understand the question and
(b) decide how to get the correct answer.

16 Katie wants to buy a car.
She decides to borrow £3500 from her father. She adds interest of 3.5% to the loan and this total is the amount she must repay her father. How much will Katie pay back to her father in total?

What does an AO3 question look like?

D AO3

Here you need to read and analyse the question. Then use your mathematical knowledge to solve this problem.

17 Rashida wishes to invest £2000 in a building society account for one year. The Internet offers two suggestions. Which of these two investments gives Rashida the greatest return?

CHESTMAN BUILDING SOCIETY
£3.50 per month
Plus **1% bonus** at the end of the year

DUNSTAN BUILDING SOCIETY
4% per annum. Paid yearly by cheque

Focus on

AO2/3

We give you extra help with AO2 and AO3 on pages 202–205.

What does a question with functional maths look like?

Functional maths is about being able to apply maths in everyday, real-life situations.

GCSE Tier	Range % of marks in the exam
Foundation	30-40
Higher	20-30

The proportion of functional maths marks in the GCSE exam depends on which tier you are taking. Don't miss out, make sure you know how to do functional maths questions!

In the exercises…

20 The Wildlife Trust are doing a survey into the number of field mice on a farm of size 240 acres. They look at one field of size 6 acres. In this field they count 35 field mice.

a Estimate how many field mice there are on the whole farm.

b Why might this be an unreliable estimate?

You need to read and understand the question. Follow your plan.

Think what maths you need and plan the order in which you'll work.

Check your calculations and make a comment if required.

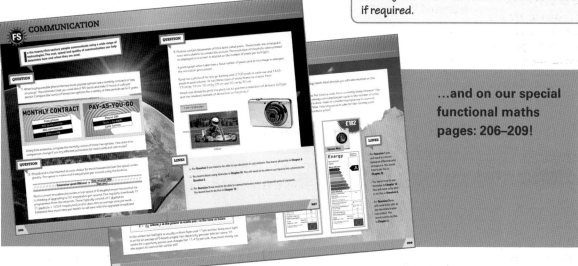

…and on our special functional maths pages: 206–209!

Quality of written communication

There will be marks in the exam for showing your working 'properly' and explaining clearly. In the exam paper, such questions will be marked with a star (*). You need to:

◉ use the correct mathematical notation and vocabulary, to show that you can communicate effectively

◉ organise the relevant information logically.

ResultsPlus

ResultsPlus features combine exam performance data with examiner insight to give you more information on how to succeed. ResultsPlus tips in the **student books** show students how to avoid errors in solutions to questions.

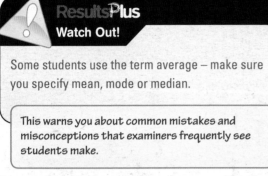

ResultsPlus
Watch Out!

Some students use the term average – make sure you specify mean, mode or median.

This warns you about common mistakes and misconceptions that examiners frequently see students make.

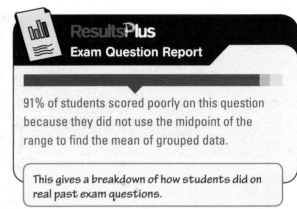

ResultsPlus
Exam Question Report

91% of students scored poorly on this question because they did not use the midpoint of the range to find the mean of grouped data.

This gives a breakdown of how students did on real past exam questions.

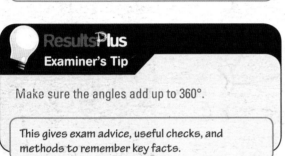

ResultsPlus
Examiner's Tip

Make sure the angles add up to 360°.

This gives exam advice, useful checks, and methods to remember key facts.

ResultsPlus in the **ActiveTeach** provides interactive practice for AO2 and AO3 questions…

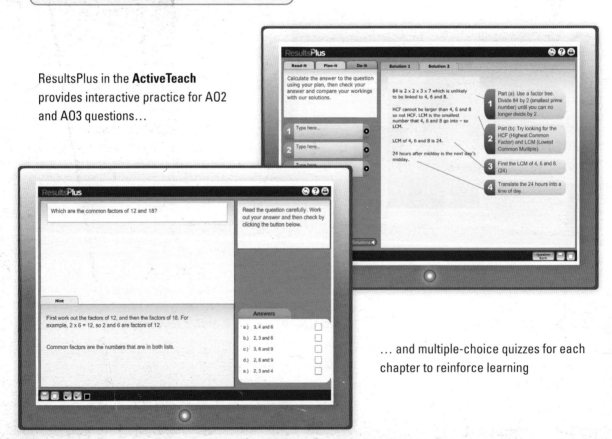

… and multiple-choice quizzes for each chapter to reinforce learning

1 NUMBER

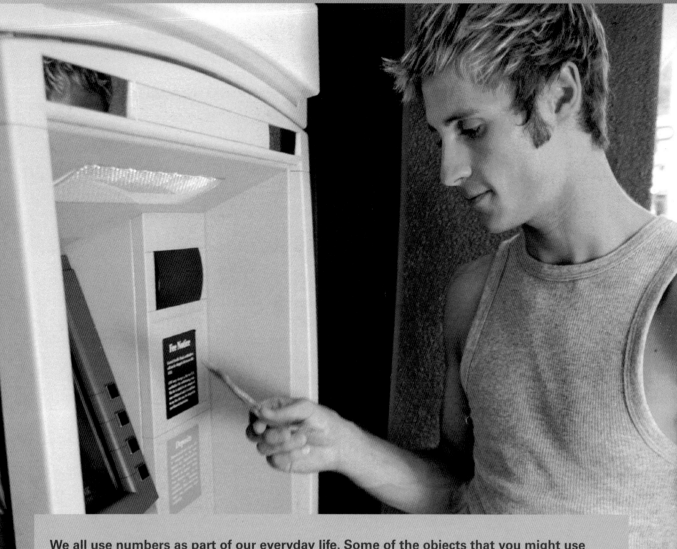

We all use numbers as part of our everyday life. Some of the objects that you might use every day such as mobile phones and cash cards need to be protected by powerful codes. Most of the codes used involve very large prime numbers.

Objectives

In this chapter you will:
- find the lowest common multiple and highest common factor of two numbers
- understand the meaning of square root and cube root
- know the correct order to carry out the different arithmetic operations
- apply the laws of indices.

Before you start

You need to:
- understand and use positive numbers and negative integers
- know how to use a number line
- know your multiplication tables
- know how to find factors and multiples of whole numbers
- be able to identify prime numbers
- understand index notation.

1.1 Understanding prime factors, LCM and HCF

Objectives

- You can express any whole number as the product of its prime factors.
- You can find the HCF of two or three numbers.
- You can find the LCM of two or three numbers.

Why do this?

If burgers come in packs of 4 and buns come in packs of 6, being able to find out the LCM of 4 and 6 is useful to make sure that there are equal numbers of burgers and buns at a barbeque.

Get Ready

1. State whether the following numbers are factors of 18, multiples of 18 or neither.
   a 4   b 6   c 9   d 36   e 12   f 3
2. State whether the following numbers are prime numbers or not.
   a 35   b 31   c 39   d 43   e 57

Key Points

- Any factor of a number that is a prime number is a **prime factor**. For example, 2 and 3 are the prime factors of 6.
- You can write any number as the product of its prime factors.
- The **Highest Common Factor (HCF)** of two whole numbers is the highest factor that is common to them both. For example, 3, 5 and 15 are all **common factors** of 30 and 45 but 15 is their highest common factor.
- The **Lowest Common Multiple (LCM)** of two whole numbers is the lowest number that is a multiple of both of them. For example, the **common multiples** of 10 and 15 are 30, 60, 90, 120, but 30 is their lowest common multiple.

Example 1 Write 120 as the product of its prime factors.

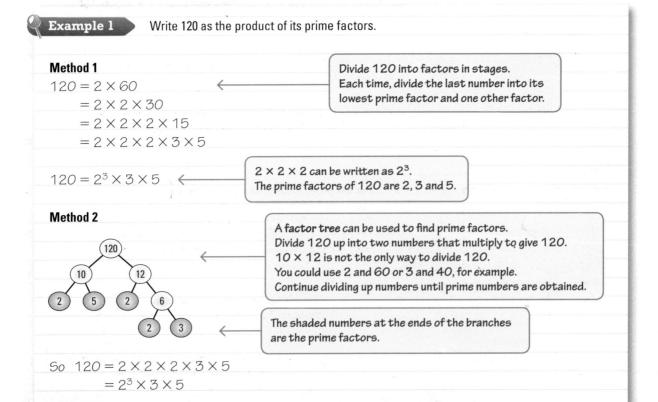

Method 1

$120 = 2 \times 60$
$\quad\; = 2 \times 2 \times 30$
$\quad\; = 2 \times 2 \times 2 \times 15$
$\quad\; = 2 \times 2 \times 2 \times 3 \times 5$

Divide 120 into factors in stages. Each time, divide the last number into its lowest prime factor and one other factor.

$120 = 2^3 \times 3 \times 5$

$2 \times 2 \times 2$ can be written as 2^3.
The prime factors of 120 are 2, 3 and 5.

Method 2

A factor tree can be used to find prime factors.
Divide 120 up into two numbers that multiply to give 120.
10×12 is not the only way to divide 120.
You could use 2 and 60 or 3 and 40, for example.
Continue dividing up numbers until prime numbers are obtained.

The shaded numbers at the ends of the branches are the prime factors.

So $120 = 2 \times 2 \times 2 \times 3 \times 5$
$\quad\quad\; = 2^3 \times 3 \times 5$

prime factor   Highest Common Factor (HCF)   common factor   Lowest Common Multiple (LCM)

Example 2   Find **a** the HCF and **b** the LCM of 6 and 10.

a The factors of 6 are 1, 2, 3, 6. ← List all the factors of 6.

The factors of 10 are 1, 2, 5, 10. ← List all the factors of 10.

The HCF of 6 and 10 is 2. ← 2 is the highest number that appears in both lists.

> The method of listing factors and multiples is best used when the given numbers are small.

b Multiples of 6 are 6, 12, 18, 24, **30**, 36 … ← List the first few multiples of 6.

Multiples of 10 are 10, 20, **30**, 40, 50, 60 … ← List the first few multiples of 10. You will need to continue listing the multiples until there is a number that appears in both lists.

The LCM of 6 and 10 is 30. ← 30 is the smallest number that appears in both lists.

Example 3   Find **a** the HCF and **b** the LCM of 140 and 210.

$140 = 2 \times 2 \times 5 \times 7$
$210 = 2 \times 3 \times 5 \times 7$ ← First express both numbers as the product of their prime factors.

Method 1

$140 = 2 \times 2 \times 5 \times 7$
$210 = 2 \times 3 \times 5 \times 7$ ← Identify the common factors; the numbers that appear in both lists.

a HCF of 140 and 210 $= 2 \times 5 \times 7$
$= 70$ ← Multiply the common factors together to get the HCF.

b LCM of 140 and 210 $= 70 \times 2 \times 3$
$= 420$ ← Multiply the HCF by the numbers in both lists that were not highlighted to get the LCM.

Method 2

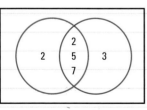

> Put the prime factors into a Venn diagram.
> The prime factors of 140 are in the blue circle.
> The prime factors of 210 are in the red circle.
> The common factors of 140 and 210 are inside the part of the diagram where the two circles intersect.

a HCF of 140 and 210 $= 2 \times 5 \times 7$
$= 70$ ← The HCF is the product of the numbers that are inside both circles.

b LCM of 140 and 210 $= 2 \times 2 \times 3 \times 5 \times 7$
$= 420$ ← The LCM is the product of all the numbers that appear in the Venn diagram.

⚙ **Exercise 1A**

C A03

* **1** Can the sum of two prime numbers be a prime number?
Explain your answer.
[*Hint:* Try adding some pairs of prime numbers.]

2 The number 48 can be written in the form $2^n \times 3$.
Find the value of n.

3 The number 84 can be written in the form $2^n \times m \times p$ where n, m and p are prime numbers.
Find the values of n, m and p.

4 Find the HCF and LCM of the following pairs of numbers.
 a 6 and 8   b 5 and 10   c 4 and 10   d 6 and 18

5 a Write 24 and 60 as products of their prime factors.
 b Find the HCF of 24 and 60.   c Find the LCM of 24 and 60.

6 a Write 72 and 120 as products of their prime factors.
 b Find the HCF of 72 and 120.   c Find the LCM of 72 and 120.

7 Find the HCF and LCM of the following pairs of numbers.
 a 36 and 90   b 54 and 72   c 60 and 96   d 144 and 180

8 $x = 2 \times 3^2 \times 5$, $y = 2^3 \times 3 \times 7$
 a Find the HCF of x and y.   b Find the LCM of x and y.

9 $m = 2^4 \times 3^2 \times 5 \times 7$, $n = 2^3 \times 5^3$
 a Find the HCF of m and n.   b Find the LCM of m and n.

B

10 Bertrand's theorem states that 'Between any two numbers n and $2n$, there always lies at least one prime number, providing n is bigger than 1'. Show that Bertrand's theorem is true:
 a for $n = 10$ b for $n = 20$ c for $n = 34$.

A03

11 A ship is at anchor between two lighthouses L and H.
The light from L shines on the ship every 30 seconds.
The light from H shines on the ship every 40 seconds.
Both lights started at the same moment.
How often do both lights shine on the ship at once?

A03

12 Burgers come in boxes of 8.
Buns come in packets of 6.
What is the smallest number of boxes of burgers and packets of buns that Mrs Moore must buy if she wants to ensure that there is a bun for every burger?

A A03

* **13** Sally says that if you multiply two prime numbers then you will always get an odd number.
Is Sally correct? Give a reason for your answer.

1.2 Understanding squares and cubes

Objectives

- You know how to find squares and cubes of whole numbers.
- You understand the meaning of square root.
- You understand the meaning of cube root.

Why do this?

If things are packed in squares you can quickly work out how many you have using square numbers, for example crates of strawberries or eggs.

Get Ready

1. Work out   **a** 6×6 **b** $2 \times 2 \times 2$ **c** -3×-3

Key Points

- A **square number** is the result of multiplying a whole number by itself.
  The square numbers can be shown as a pattern of squares.

$1^2 = 1 \times 1 = 1$	$2^2 = 2 \times 2 = 4$	$3^2 = 3 \times 3 = 9$	$4^2 = 4 \times 4 = 16$
1st square number	2nd square number	3rd square number	4th square number

- A **cube number** is the result of multiplying a whole number by itself then multiplying by that number again.
  The cube numbers can be shown as a pattern of cubes.

$1^3 = 1 \times 1 \times 1 = 1$	$2^3 = 2 \times 2 \times 2 = 8$	$3^3 = 3 \times 3 \times 3 = 27$
1st cube number	2nd cube number	3rd cube number

- To find the **square** of any number, multiply the number by itself.
  The square of $-4 = (-4)^2 = -4 \times -4 = 16$.
- $5 \times 5 = 25$, so we say that 5 is the **square root** of 25. It is a number that when multiplied by itself gives 25.
  You can write the square root of 25 as $\sqrt{25}$. The square root of 25 can also be -5 because $-5 \times -5 = 25$.
- To find the **cube** of any number, multiply the number by itself then multiply by the number again.
  The cube of $-2 = (-2)^3 = -2 \times -2 \times -2 = -8$.
- $-2 \times -2 \times -2 = -8$, so we say that -2 is the **cube root** of -8. It is a number that when multiplied by itself, then multiplied by itself again, gives -8. You can write the cube root of -8 as $\sqrt[3]{-8}$.

Example 4

Find   **a** the 6th square number
       b the 10th cube number.

a The 6th square number is $6^2 = 6 \times 6$
$= 36$

b The 10th cube number is $10^3 = 10 \times 10 \times 10$
$= 1000$

ResultsPlus
Examiner's Tip

You need to know the integer squares and corresponding square roots up to 15×15, and the cubes of 2, 3, 4, 5 and 10.

Exercise 1B

1 Write down:
   a the first 15 square numbers
   b the first 5 cube numbers.

2 From each list write down all the numbers which are:
   i square numbers   **ii** cube numbers.

   a 50, 20, 64, 30, 1, 80, 8, 49, 9
   b 10, 21, 57, 4, 60, 125, 7, 27, 48, 16, 90, 35
   c 137, 150, 75, 110, 50, 125, 64, 81, 144
   d 90, 180, 125, 100, 81, 75, 140, 169, 64

> **ResultsPlus**
> **Examiner's Tip**
>
> You need to be able to recall:
> • integer squares from 2×2 up to 15×15 and the corresponding square roots
> • the cubes of 2, 3, 4, 5 and 10.

Example 5  Find   **a** $(-3)^2$ **b** $\sqrt{100}$ **c** $(-4)^3 + \sqrt[3]{125}$

a $(-3)^2 = -3 \times -3 = 9$ ← Two signs the same so answer is positive.

b $\sqrt{100} = 10$

c $(-4)^3 = -4 \times -4 \times -4 = -64$
   $\sqrt[3]{125} = 5$ ← $5^3 = 125$ so the cube root of 125 is 5.
   $(-4)^3 + \sqrt[3]{125} = -64 + 5$
   $= -59$

> **ResultsPlus**
> **Examiner's Tip**
>
> Remember, when multiplying or dividing:
> two signs the same give a $+$
> two different signs give a $-$.

Exercise 1C

1 Work out
   a 3^2 **b** 7^2 **c** 4^3 **d** 10^3 **e** 11^2

2 Write down
   a $\sqrt{36}$ **b** $\sqrt{16}$ **c** $\sqrt{81}$ **d** $\sqrt{1}$ **e** $\sqrt{64}$

3 Work out
   a $(-6)^2$ **b** $(-2)^3$ **c** $(-9)^2$ **d** $(-1)^3$ **e** $(-12)^2$

4 Write down
   a $\sqrt[3]{8}$ **b** $\sqrt[3]{-27}$ **c** $\sqrt[3]{-1}$ **d** $\sqrt[3]{64}$ **e** $\sqrt[3]{1000}$

5 Work out
   a $3^2 + 2^3$ **b** $\sqrt{4} \times 5^2$ **c** $5^2 \times \sqrt{100}$ **d** $\sqrt[3]{-8} + 4^2$
   e $\sqrt[3]{1000} \div \sqrt{100}$ **f** $4^3 \div 2^3$ **g** $(-1)^3 + 2^3 - (-3)^3$ **h** $4^2 + (-3)^3$
   i $\dfrac{6^2}{2^2}$ **j** $5^2 \times \dfrac{\sqrt{16}}{\sqrt[3]{8}}$ **k** $2^3 \times \dfrac{\sqrt{100}}{\sqrt{64}}$ **l** $\dfrac{4^2 - \sqrt[3]{-8}}{\sqrt{9}}$

1.3 Understanding the order of operations

⊙ Objective

⊙ You know and can apply the order of operations.

⬆ Get Ready

1. Work out   **a** 6×3 **b** 4^2 **c** $70 \div 7$

⬡ Why do this?

When following a recipe, you need to add the ingredients in the right order. The same is true of calculations such as $3 \times 4 + 2 \times 5$. The operations must be carried out in the correct order or the answer will be wrong.

Key Points

◉ **BIDMAS** gives the order in which each **operation** should be carried out.

◉ Remember that **B I D M A S** stands for:

B rackets	If there are brackets, work out the **value** of the expression inside the brackets first.
I ndices	Indices include square roots, cube roots and **powers**.
D ivide	If there are no brackets, do dividing and multiplying before adding and subtracting, no matter where they come in the expression.
M ultiply	
A dd	
S ubtract	If an expression has only adding and subtracting then work it out from left to right.

Example 6

$$10 \times 2^2 - 5 \times 3 = 10 \times 4 - 5 \times 3$$
$$= 40 - 15$$
$$= 25$$

Work out 2^2 first, then do all the multiplying before the subtraction.

Example 7

$$(12 - 2 \times 5)^3 = (12 - 10)^3$$
$$= 2^3$$
$$= 8$$

The sum in the bracket is worked out first. Work out 2×5 and then do the subtraction.

⚙ Exercise 1D

1 Work out

a $5 \times (2 + 3)$	**b** $5 \times 2 + 3$	**c** $20 \div 4 + 1$	**d** $20 \div (4 + 1)$
e $(6 + 4) \div -2$	**f** $6 + 4 \div 2$	**g** $24 \div (6 - 2)$	**h** $24 \div 6 - 2$
i $7 - (4 + 2)$	**j** $7 - 4 + 2$	**k** $5 \times 4 - 2 \times 3$	**l** $28 - 4 \times -6$
m $14 + 3 \times 6$	**n** $6 + 3 \times 5 - 12 \div 2$	**o** $25 - 5 \times 4 + 3$	**p** $(15 - 5) \times (4 + 3)$

2 Work out

a $(3 + 4)^2$	**b** $3^2 + 4^2$	**c** $3 \times (4 + 5)^2$	**d** $3 \times 4^2 + 3 \times 5^2$
e $2 \times (4 + 2)^2$	**f** $3 \times \sqrt{25} + 2 \times 3^3$	**g** $\dfrac{(2 + 5)^2}{3^2 - 2}$	**h** $\dfrac{5^2 - 2^2}{-3}$

D

B

3 **Work out**

a $(2 + 3)^3 \div \sqrt{25}$

b $((15 - 5) \times 4) \div ((2 + 3) \times 2)$

c $2^3 + 6^2 \div \sqrt{9} - 4 \times 3$

d $(\sqrt[3]{-27} - 2)^2 + \sqrt{3^2 \times 2^2}$

1.4 Understanding the index laws

Objectives

○ You can use index notation.

○ You can use index laws.

Why do this?

Using the index laws you can work out that you have $2^5 = 32$ great great great grandparents.

Get Ready

1. Work out 2^5

2. Work out 5^3

3. Work out $27^4 \div 27^2$

Key Points

◉ A number written in the form a^n is an **index number**.

◉ The **laws of indices** are:

$a^m \times a^n = a^{m + n}$ To multiply two powers of the same number add the indices.

$a^m \div a^n = a^{m - n}$ To divide two powers of the same number subtract the indices.

$(a^m)^n = a^{m \times n}$ To raise a power to a further power multiply the indices together.

You will encounter negative and fractional indices in Chapter 5.

Example 8   Work out   **a** 3^4 **b** 2^6

a $3^4 = 3 \times 3 \times 3 \times 3$

    $= 81$

b $2^6 = 2 \times 2 \times 2 \times 2 \times 2 \times 2$

    $= 64$

ResultsPlus

Watch Out!

Remember that a^3 means that you multiply three as together. It does not mean $a \times 3$.

Example 9   Write each expression as a power of 5.   **a** $5^6 \times 5^4$ **b** $5^{12} \div 5^4$ **c** $(5^3)^2$

a $5^6 \times 5^4 = 5^{4+6}$ ← Use the index law $a^m \times a^n = a^{m+n}$

    $= 5^{10}$

b $5^{12} \div 5^4 = 5^{12-4}$ ← Use the index law $a^m \div a^n = a^{m-n}$

    $= 5^8$

c $(5^3)^2 = 5^{3 \times 2}$ ← Use the index law $(a^m)^n = a^{m \times n}$

    $= 5^6$

Example 10 Work out $\dfrac{4^7 \times 4}{4^5}$

$\dfrac{4^7 \times 4}{4^5} = \dfrac{4^7 \times 4^1}{4^5}$

$= \dfrac{4^8}{4^5}$ ← Simplify the top of the fraction, add 7 and 1.

$= 4^3$

$= 4 \times 4 \times 4 = 64$ ← As the question asks you to 'Work out', the final answer must be a number.

ResultsPlus
Examiner's Tip
'Work out' means 'evaluate' the expression, rather than leaving the answer as a power.

ResultsPlus
Watch Out!
Remember that a is the same as a^1.

Exercise 1E

1 Write as a power of a single number
a $6^5 \times 6^7$ b $4^7 \div 4^2$ c $(7^2)^3$ d $5^9 \div 5^3$ e $3^8 \times 3^2$

2 Work out
a $10^2 \times 10^3$ b $5^7 \div 5^4$ c $(2^3)^2$ d $3^4 \div 3^2$ e 4×4^2

3 Find the value of n
a $3^n \div 3^2 = 3^3$ b $8^5 \div 8^n = 8^2$ c $2^5 \times 2^n = 2^{10}$ d $3^n \times 3^5 = 3^9$ e $2^6 \times 2^3 = 2^n$

4 Write as a power of a single number
a $\dfrac{3^3 \times 3^5}{3^4}$ b $\dfrac{5^6 \times 5^7}{5^4}$ c $\dfrac{2^8 \times 2^5}{2^7}$ d $\dfrac{6^{15}}{6 \times 6^9}$ e $\dfrac{4^2 \times 4^7}{4^3 \times 4^4}$

5 Work out
a $\dfrac{3^3 \times 3^5}{3^6}$ b $\dfrac{2^6 \times 2^2}{2^4}$ c $\dfrac{4^7}{4 \times 4^4}$ d $\dfrac{10^5 \times 10^6}{10^7}$ e $\dfrac{7^8 \times 7}{7^3 \times 7^4}$

6 Work out the value of n in the following
a $40 = 5 \times 2^n$ b $32 = 2^n$ c $20 = 2^n \times 5$ d $48 = 3 \times 2^n$ e $54 = 2 \times 3^n$

Chapter review

Key Points

- Any factor of a number that is a prime number is a **prime factor**.
- You can write any number as the product of its prime factors.
- The **Highest Common Factor (HCF)** of two whole numbers is the highest factor that is common to them both.
- The **Lowest Common Multiple (LCM)** of two whole numbers is the lowest number that is a multiple of both of them.
- A **square number** is the result of multiplying a whole number by itself.
- A **cube number** is the result of multiplying a whole number by itself then multiplying by that number again.

- To find the **square** of any number, multiply the number by itself.
- The **square root** of 25 is a number that when multiplied by itself gives 25.
  You can write the square root of 25 as $\sqrt{25}$.
  The square root of 25 can also be -5 because $-5 \times -5 = 25$.
- To find the **cube** of any number, multiply the number by itself then multiply by the number again.
- The **cube root** of -8 is a number that when multiplied by itself, then multiplied by itself again, gives -8.
  You can write the cube root of -8 as $\sqrt[3]{-8}$.
- **BIDMAS** gives the order in which **operations** should be carried out.
- Remember that **BIDMAS** stands for:

  Brackets   If there are brackets, work out the **value** of the expression inside the brackets first.

  Indices   Indices include square roots, cube roots and **powers**.

  Divide   If there are no brackets, do dividing and multiplying before adding and subtracting, no

  Multiply   matter where they come in the expression.

  Add   If an expression has only adding and subtracting then work it out from left to right.

  Subtract

- A number written in the form a^n is an **index number**.
- The **laws of indices** are:

  $a^m \times a^n = a^{m+n}$ To multiply two powers of the same number add the indices.

  $a^m \div a^n = a^{m-n}$ To divide two powers of the same number subtract the indices.

  $(a^m)^n = a^{m \times n}$ To raise a power to a further power multiply the indices together.

Review exercise

A03

1 Jim writes down the numbers from 1 to 100. Ben puts a red spot on all the even numbers and Helen puts a blue spot on all the multiples of 3.

   a What is the largest number that has both a red and a blue spot?

   b How many numbers have neither a blue nor a red spot?

   Sophie puts a green spot on all the multiples of 5.

   c How many numbers have exactly two coloured spots on them?

D

2 Find the missing numbers in each case.

   a $2 \times ? + (-3) = (-7)$ **b** $(-4) \times ? + 5 = (-3)$ **c** $? \div 2 + 4 = (-4)$

A02 A03

3 Neal works part time in a local supermarket, stacking shelves.

He has been asked to use the pattern below to advertise a new brand of beans.

This stack is 3 cans high.

   a How many cans will he need to build a stack 10 cans high?

   b If he has been given 200 cans, how many cans high would his stack be?

Next he is asked to stack cans of tomato soup in a similar shape, but this time it is two cans deep.

Use your answers to parts **a** and **b** to answer the following questions.

c How many cans will he need to build a stack 10 cans high?

d If he has been given 400 cans, how many cans high would his stack be?

4 A chocolate company wishes to produce a presentation box of 36 chocolates for Valentine's Day.
It decides that a rectangular shaped box is the most efficient, but needs to decide how to arrange the chocolates.

How many different possible arrangements are there:

a using one layer

b using two layers

c using three layers.

Which one do you think would look best?

5 The number 1 is a square number and a cube number. Find another number which is a square number and a cube number.

6 $4^2 \times 6^2 = 576$

Work out **a** $40^2 \times 60^2$ **b** $400^2 \times 6^2$ **c** $5760 \div 6^2$ **d** $4^2 \times 60^2$ **e** $4^3 \times 6^2$

7 Work out **a** $2 + 4 \div 4$ **b** $5^3 \div 5 + 5$ **c** $(2^2)^3 - (2^3)^2$

8 Simplify **a** $\dfrac{3^5 \times 3^3}{3^6}$ **b** $\dfrac{4^4 \times 4^7}{4^{10}}$ **c** $(2^4)^3$ **d** $\dfrac{5^{12}}{5^7 \times 5^3}$

9 **a** Express 252 as a product of its prime factors.

b Express 6×252 as a product of prime factors.

10 James thinks of two numbers.
He says 'The highest common factor (HCF) of my two numbers is 3.
The lowest common multiple (LCM) of my two numbers is 45'.
Write down the two numbers James could be thinking of.

ResultsPlus
Exam Question Report

75% of students answered this sort of question well because they chose the right method to answer the question.

June 2008

11 Write 84 as a product of its prime factors.
Hence or otherwise write 168^2 as a product of its prime factors.

C **A03** 12 A car's service book states that the air filter must be replaced every 10 000 miles and the
diesel fuel filter every 24 000 miles.
After how many miles will both need replacing at the same time?

B 13 Work out

a $\dfrac{\sqrt[2]{81}}{3} \times 4^2$ b $(\sqrt[3]{216})^2$ c $(\sqrt{49})^3$ d $\dfrac{7^2 + \sqrt[3]{1}}{\sqrt[3]{8}}$

A03 14 $2^{30} \div 8^9 = 2^x$

Work out the value of x.   *Nov 2007*

A* **A03** 15 Write whether each of the following statements is true or false. If the statement is false give an
example to show it.

a The sum of two prime numbers is always a prime number.

b The sum of two square numbers is never a prime number.

c The difference between consecutive prime numbers is never 2.

d The product of two prime numbers is always a prime number.

e No prime number is a square number.

16 a Take a piece of scrap A4 paper.

If you fold it in half you create two equal pieces. Fold it in half again; you now have four equal pieces.

It is said that no matter how large and how thin you make the paper, it cannot be folded more than
seven times. Try it.

If you fold it seven times, how many equal pieces does the paper now have?

b In 2001, there were two rabbits left on an island.

A simple growth model predicts that in 2002 there will be four rabbits and in 2003, eight rabbits.

The population of rabbits continues to double every year.

How long is it before there are 1 million rabbits on the island?

2 FRACTIONS

Only one eighth of an iceberg shows above the surface of the water, which leaves most of it hidden. The largest northern hemisphere iceberg was encountered near Baffin Island in Canada in 1882. It was 13 km long, 6 km wide and had a height above water of about 20 m. It had a mass of over 9 billion tonnes – enough water for everyone in the world to drink a litre a day for over four years.

Objectives

In this chapter you will:
- add, subtract, multiply and divide fractions and mixed numbers
- find a fraction of a quantity
- solve problems involving fractions.

Before you start

You need to be able to:
- find the highest common factor (HCF) of two numbers
- find the lowest common multiple (LCM) of two numbers
- simplify and order fractions
- convert between improper fractions and mixed numbers.

2.1 Adding and subtracting fractions and mixed numbers

⊙ Objectives

○ You can add and subtract fractions.
○ You can add and subtract mixed numbers.

⊘ Why do this?

Measurements are not always given in whole numbers. You may need to find the total length of two distances given as fractions, for example, $2\frac{3}{4}$ km and $1\frac{1}{4}$ km.

⊕ Get Ready

1. Write $\frac{32}{36}$ in its simplest form.　**2.** Change $2\frac{3}{8}$ to an improper fraction.　**3.** Change $\frac{47}{5}$ to a mixed number.

🔍 Key Points

○ To add (or subtract) fractions, change them to **equivalent fractions** that have the same denominator. This new demonimator will be the LCM of the two denominators (see Section 1.1 for LCM). Then add (or subtract) the numerators but do not change the denominator.
○ To add (or subtract) **mixed numbers**, add (or subtract) the whole numbers, then add (or subtract) the fractions separately.

Example 1 Work out $\frac{7}{8} - \frac{1}{4}$

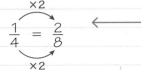

The LCM of 4 and 8 is 8.
Convert $\frac{1}{4}$ to the equivalent fraction with a denominator of 8.

$$\frac{7}{8} - \frac{1}{4} = \frac{7}{8} - \frac{2}{8}$$

Subtract the numerators only.

$$= \frac{5}{8}$$

Results Plus
Examiner's Tip

If you could use several numbers as the new denominator, using the LCM of the two denominators means you won't need to simplify later.

Example 2 Work out $\frac{5}{6} + \frac{7}{10}$
Give your answer as a mixed number.

$$\frac{5}{6} \xrightarrow{\times 5} = \frac{25}{30} \qquad \frac{7}{10} \xrightarrow{\times 3} = \frac{21}{30}$$

The LCM of 6 and 10 is 30.
Convert each fraction to its equivalent fraction with a denominator of 30.

$$\frac{5}{6} + \frac{7}{10} = \frac{25}{30} + \frac{21}{30}$$

Simplify the fraction.
Divide each number in the fraction by 2.

$$= \frac{46}{30}$$

$$= \frac{23}{15}$$

Convert the mixed number to an improper fraction.
$23 \div 15 = 1$ remainder 8.

$$= 1\frac{8}{15}$$

Exercise 2A

Questions in this chapter are targeted at the grades indicated.

Give each answer as a fraction in its simplest form.

1 Work out

a $\frac{5}{11} + \frac{3}{11}$ b $\frac{1}{9} + \frac{4}{9}$ c $\frac{7}{15} + \frac{4}{15}$ d $\frac{3}{10} + \frac{1}{10}$

2 Work out

a $\frac{1}{5} + \frac{1}{2}$ b $\frac{1}{3} + \frac{1}{7}$ c $\frac{3}{7} + \frac{2}{5}$ d $\frac{5}{9} + \frac{1}{3}$

e $\frac{2}{5} + \frac{1}{4}$ f $\frac{1}{2} + \frac{2}{9}$ g $\frac{2}{9} + \frac{1}{6}$ h $\frac{7}{20} + \frac{2}{5}$

3 Work out

a $\frac{1}{2} - \frac{1}{4}$ b $\frac{1}{3} - \frac{1}{4}$ c $\frac{7}{8} - \frac{2}{5}$ d $\frac{8}{9} - \frac{2}{3}$

e $\frac{3}{4} - \frac{1}{2}$ f $\frac{2}{3} - \frac{1}{4}$ g $\frac{17}{20} - \frac{3}{4}$ h $\frac{5}{6} - \frac{7}{9}$

Give each answer as a fraction or a mixed number in its simplest form.

4 Work out

a $\frac{2}{5} + \frac{7}{8}$ b $\frac{3}{4} + \frac{4}{5}$ c $\frac{5}{6} - \frac{2}{3}$ d $\frac{7}{10} + \frac{1}{4}$

e $\frac{5}{6} + \frac{9}{10}$ f $\frac{3}{4} + \frac{7}{8}$ g $\frac{4}{5} - \frac{1}{2} + \frac{9}{10}$ h $\frac{7}{9} + \frac{2}{3} - \frac{1}{6}$

D

Example 3   Work out $5\frac{7}{10} + 4\frac{1}{2}$

$5\frac{7}{10} + 4\frac{1}{2} = 9\frac{7}{10} + \frac{1}{2}$ ← Add the whole numbers.

$= 9\frac{7}{10} + \frac{5}{10}$ ← Convert the fractions into equivalent fractions with a denominator of 10.

$= 9\frac{12}{10}$ ← $\frac{12}{10}$ is an improper fraction. Change this into a mixed number.

$= 10\frac{2}{10}$ ← Simplify $\frac{2}{10}$.

$= 10\frac{1}{5}$

Exercise 2B

1 Work out   a $6\frac{3}{4} + 1\frac{1}{2}$ b $4\frac{4}{5} + \frac{1}{2}$ c $7\frac{5}{6} + 3\frac{2}{7}$ d $12\frac{3}{4} + 5\frac{2}{5}$

C

2 Becky cycled $2\frac{3}{4}$ miles to one village then a further $4\frac{1}{3}$ miles to her home. What is the total distance that Becky cycled?

3 A bag weighs $\frac{3}{7}$ lb. The contents weigh $1\frac{1}{5}$ lb. What is the total weight of the bag and its contents?

Example 4 Work out $7\frac{1}{4} - 2\frac{7}{10}$

Method 1

$$7\frac{1}{4} - 2\frac{7}{10} = 5\frac{5}{20} - \frac{14}{20}$$

> $\frac{5}{20} - \frac{14}{20}$ will give a negative result.
> Write $5\frac{5}{20}$ as $4 + 1\frac{5}{20} = 4\frac{25}{20}$.

$$= 4\frac{25}{20} - \frac{14}{20}$$

$$= 4\frac{11}{20}$$

Method 2

$$7\frac{1}{4} - 2\frac{7}{10} = \frac{29}{4} - \frac{27}{10}$$

> Convert the mixed numbers to improper fractions.

$$= \frac{290}{40} - \frac{108}{40}$$

$$= \frac{182}{40}$$

$$= 4\frac{22}{40}$$

> Convert the improper fraction to a mixed number.
> $182 \div 40 = 4$, remainder 22.

$$= 4\frac{11}{20}$$

> Simplify the fraction.

Exercise 2C

1 Work out   **a** $2\frac{1}{2} - 1\frac{1}{4}$ **b** $3\frac{7}{8} - 1\frac{1}{2}$ **c** $6 - 5\frac{1}{4}$ **d** $8 - 4\frac{2}{3}$

2 Work out   **a** $2\frac{1}{4} - 1\frac{1}{2}$ **b** $3\frac{1}{4} - 1\frac{2}{3}$ **c** $4\frac{2}{7} - 1\frac{3}{5}$ **d** $7\frac{1}{9} - 3\frac{2}{3}$

3 A box containing vegetables has a total weight of $5\frac{1}{4}$ kg. The empty box has a weight of $1\frac{7}{8}$ kg. What is the weight of the vegetables?

4 A tin contains $7\frac{1}{2}$ pints of oil. Julie pours out $4\frac{5}{8}$ pints from the tin. How much oil remains?

2.2 Multiplying fractions and mixed numbers

Objectives

- You can multiply fractions.
- You can multiply mixed numbers.
- You can find a fraction of a quantity.

Why do this?

Shops often advertise discounts as '$\frac{2}{3}$ off the normal price'. To work out the discount you will need to multiply by $\frac{2}{3}$.

Get Ready

1. Work out 4×8.

2. Work out 5×9.

3. Change $3\frac{4}{5}$ to an improper fraction.

4. Convert $\frac{14}{3}$ to a mixed number.

> **Key Points**

- To multiply fractions:
  - Convert any mixed numbers to **improper fractions**.
  - Simplify if possible.
  - Multiply the numerators and multiply the denominators.

Example 5   Work out $\frac{2}{3} \times 7$.

$$\frac{2}{3} \times 7 = \frac{2}{3} \times \frac{7}{1}$$ ← Write 7 as an improper fraction.

$$= \frac{2 \times 7}{3 \times 1}$$

$$= \frac{14}{3}$$

$$= 4\frac{2}{3}$$ ← Write $\frac{14}{3}$ as a mixed number.

Example 6   Work out $\frac{5}{6}$ of 9 metres.

$$\frac{5}{6} \times 9 = \frac{5}{6} \times \frac{9}{1}$$ ← To find the fraction of a quantity, multiply the fraction by the quantity.

$$= \frac{5 \times \cancel{9}^3}{_2\cancel{6} \times 1}$$ ← Simplify by dividing the numerator and denominator by 3.

$$= \frac{15}{2}$$

$$= 7\frac{1}{2} \text{ metres}$$

Example 7   Work out $\frac{5}{14} \times \frac{7}{10}$

$$\frac{5}{14} \times \frac{7}{10} = \frac{5 \times 7}{14 \times 10}$$

$$= \frac{{}^1\cancel{5} \times 7}{14 \times \cancel{10}_2}$$ ← Simplify by dividing the numerator and denominator by 5.

$$= \frac{1 \times \cancel{7}^1}{_2\cancel{14} \times 2}$$ ← Simplify by dividing the numerator and denominator by 7.

$$= \frac{1}{4}$$

Example 8   Work out $2\frac{2}{3} \times 1\frac{4}{5}$

$$2\frac{2}{3} \times 1\frac{4}{5} = \frac{8}{3} \times \frac{9}{5}$$ ← Convert each mixed number into an improper fraction.

$$= \frac{8 \times \cancel{9}^3}{_1\cancel{3} \times 5}$$ ← Divide the numerator and denominator by 3.

$$= \frac{24}{5}$$

$$= 4\frac{4}{5}$$ ← Convert the improper fraction into a mixed number.

Exercise 2D

1 Work out

a $\frac{3}{5} \times \frac{1}{2}$ b $\frac{1}{4} \times \frac{3}{5}$ c $\frac{10}{11} \times \frac{3}{5}$ d $\frac{5}{6} \times \frac{4}{15}$

e $\frac{2}{3} \times \frac{2}{7}$ f $\frac{3}{4} \times \frac{3}{5}$ g $\frac{9}{28} \times \frac{14}{15}$ h $\frac{25}{36} \times \frac{27}{40}$

2 Work out

a $2 \times \frac{1}{3}$ b $3 \times \frac{1}{4}$ c $\frac{9}{20} \times 8$ d $\frac{3}{5} \times 25$

3 Work out

a $\frac{3}{5}$ of 35 kg   b $\frac{4}{9}$ of 15 m   c $\frac{5}{8}$ of 12 litres   d $\frac{3}{10}$ of 25 pints

4 Jomo delivers 56 newspapers on his round. On Fridays $\frac{3}{8}$ of the newspapers have a magazine supplement. How many supplements does he deliver?

5 Barry earns £130.60 in one week. He pays $\frac{1}{4}$ of this in tax. How much money does he pay in tax each week?

6 Work out

a $1\frac{1}{4} \times \frac{1}{3}$ b $1\frac{3}{5} \times \frac{1}{2}$ c $3\frac{3}{4} \times 1\frac{1}{10}$ d $1\frac{2}{3} \times 4\frac{1}{5}$

e $1\frac{1}{3} \times 2\frac{1}{4}$ f $3\frac{1}{2} \times 1\frac{1}{4}$ g $6\frac{3}{7} \times 1\frac{5}{9}$ h $8\frac{1}{3} \times 2\frac{7}{10}$

7 Kieran takes $2\frac{1}{4}$ minutes to complete one lap at the Go Kart Centre. How long will it take him to complete $6\frac{1}{2}$ laps?

8 A melon weighs $2\frac{1}{2}$ lb. Work out the weight of $8\frac{1}{4}$ melons.

2.3 Dividing fractions and mixed numbers

◎ Objectives

○ You can divide fractions.
○ You can divide mixed numbers.

❓ Why do this?

A carpenter may want to work out how many pieces of wood measuring $\frac{3}{4}$ m he can cut from a 5 m piece of wood.

⬆ Get Ready

1. Work out $\frac{3}{7} \times \frac{2}{5}$.
2. Work out $\frac{2}{3} \times \frac{1}{4}$.
3. Write $3\frac{3}{7}$ as an improper fraction.

Key Points

◉ To divide fractions:
 ◉ Convert any mixed numbers to improper fractions.
 ◉ Convert divide to multiply and invert the second fraction (**inverted** means turned upside down).
 ◉ Multiply the numerators and multiply the denominators.

inverted

Example 9 Work out $\frac{4}{5} \div 3$

Multiplying by $\frac{1}{3}$ is the same as dividing by 3. $\frac{1}{3}$ is called the reciprocal of 3.

$\frac{4}{5} \div 3 = \frac{4}{5} \div \frac{3}{1}$ ← Write the whole number as an improper fraction.

$= \frac{4}{5} \times \frac{1}{3}$ ← Change ÷ to × and turn the second fraction upside down. Multiply the fractions.

$= \frac{4}{15}$

Example 10 Work out $\frac{5}{6} \div \frac{3}{4}$. Give your answer in its simplest form.

$\frac{5}{6} \div \frac{3}{4} = \frac{5}{6} \times \frac{4}{3}$ ← Turn $\frac{3}{4}$ upside down to get $\frac{4}{3}$.

$= \frac{5}{\cancel{6}_3} \times \frac{\cancel{4}^2}{3}$ ← Divide the numerator and denominator by 2.

$= \frac{10}{9}$

$= 1\frac{1}{9}$ ← Write the improper fraction as a mixed number.

Example 11 Work out $2\frac{4}{5} \div 2\frac{1}{10}$

$2\frac{4}{5} \div 2\frac{1}{10} = \frac{14}{5} \div \frac{21}{10}$ ← Write the mixed numbers as improper fractions.

$= \frac{14}{5} \times \frac{10}{21}$ ← Turn $\frac{21}{10}$ upside down to get $\frac{10}{21}$.

$= \frac{\cancel{14}^2}{\cancel{5}_1} \times \frac{\cancel{10}^2}{\cancel{21}^3}$ ← Divide top and bottom by 7 and by 2.

$= \frac{4}{3}$

$= 1\frac{1}{3}$ ← Write the improper fraction as a mixed number.

Exercise 2E

1 Work out

a $\frac{5}{6} \div 2$ b $\frac{3}{8} \div 2$ c $\frac{4}{5} \div \frac{3}{10}$ d $\frac{9}{16} \div \frac{3}{8}$

e $\frac{1}{4} \div \frac{1}{3}$ f $\frac{3}{5} \div \frac{1}{2}$ g $\frac{20}{21} \div \frac{8}{15}$ h $\frac{25}{32} \div \frac{15}{16}$

2 Work out

a $3\frac{1}{2} \div 7$ b $2\frac{4}{5} \div \frac{1}{10}$ c $3\frac{3}{4} \div 1\frac{4}{5}$ d $6\frac{2}{3} \div 2\frac{8}{9}$

e $1\frac{1}{2} \div \frac{3}{4}$ f $2\frac{4}{9} \div \frac{2}{3}$ g $7\frac{1}{2} \div 1\frac{1}{4}$ h $2\frac{1}{12} \div 1\frac{1}{9}$

3 A tin holds $10\frac{2}{3}$ litres of methylated spirit for a lamp. How many times will it fill a lamp holding $\frac{2}{3}$ litre?

4 A metal rod is $10\frac{4}{5}$ metres long. How many short rods $\frac{3}{10}$ metre long can be cut from the longer rod?

5 Tar and Stone can resurface $2\frac{1}{5}$ km of road in a day. How many days will it take them to resurface a road of length $24\frac{3}{5}$ km?

2.4 Solving fraction problems

◉ Objective

○ You can solve problems involving fractions.

◈ Why do this?

A vet may need to work out fractions of a dosage depending on the size of the animal in comparison to the standard.

⬆ Get Ready

1. Work out $\frac{3}{4} + \frac{1}{8}$

2. Work out $\frac{4}{9} \times \frac{1}{5}$

3. Work out $5\frac{7}{8} - 2\frac{1}{4}$

🔍 Key Point

◉ You can use your knowledge of fractions to solve problems from real life.

Example 12

In a cinema $\frac{2}{5}$ of the audience are women, $\frac{1}{8}$ of the audience are men.
All the rest of the audience are children.
What fraction of the audience are children?

$$\frac{2}{5} + \frac{1}{8} = \frac{16}{40} + \frac{5}{40}$$

Add $\frac{2}{5}$ and $\frac{1}{8}$ to find the fraction of the audience who are women or men.

$$= \frac{21}{40}$$

$$1 - \frac{21}{40} = \frac{40}{40} - \frac{21}{40}$$

Subtract $\frac{21}{40}$ from 1 to find the fraction of the audience who are children.

$$= \frac{19}{40}$$

$\frac{19}{40}$ of the audience are children.

Example 13

A school has 1800 pupils. 840 of these pupils are girls.
$\frac{3}{4}$ of the girls like swimming. $\frac{1}{3}$ of the boys like swimming.
Work out the total number of pupils in the school who like swimming.

$$\frac{3}{4} \times 840 = 630$$

Work out the number of girls who like swimming.

$$1800 - 840 = 960$$

Work out the number of boys in the school.

$$\frac{1}{3} \times 960 = 320$$

Work out the number of boys who like swimming.

$$630 + 320 = 950$$

Work out the total number of pupils who like swimming.

950 pupils like swimming.

Exercise 2F

1 Simon spends $\frac{1}{2}$ of his money on rent and $\frac{1}{3}$ of his money on transport.
  a What fraction of his money does he spend on rent and transport altogether?
  b What fraction of his money is left?

2 $\frac{8}{9}$ of an iceberg lies below the surface of the water. The total volume of an iceberg is 990 m³.
  What volume of this iceberg is below the surface?

3 DVDs are sold for £14 each. $\frac{2}{5}$ of the £14 goes to the DVD company.
  How much of the £14 goes to the DVD company?

4 An MP3 player usually costs £130. In a sale all prices are reduced by $\frac{2}{5}$.
  Work out the sale price of the MP3 player.

5 A factory has 1710 workers. 650 of the workers are female.
  $\frac{2}{5}$ of the female workers are under the age of 30, $\frac{1}{4}$ of the male workers are under the age of 30.
  How many workers in total are aged under 30?

6 There are 36 students in a class. Javed says that $\frac{3}{8}$ of these students are boys.
  Explain why Javed cannot be right.

7 Tammy watches two films. The first film is $1\frac{3}{4}$ hours long and the second one is $2\frac{1}{3}$ hours long.
  Work out the total length of the two films.

8 $\frac{2}{3}$ of a square is shaded. $\frac{3}{4}$ of the shaded part is shaded blue.
  What fraction of the whole square is shaded blue?

9 Alison, Becky and Carol take part in a charity relay race. The race is over a total distance of $2\frac{5}{8}$ km.
  Each girl runs an equal distance. Work out how far each girl runs.

10 In a book, $\frac{3}{8}$ of the pages have pictures on them.
  Given that 72 pages have a picture on, work out the number of pages in the book.

11 Jed buys some oranges. He sells $\frac{3}{5}$ of these oranges.
  Of the oranges he has left, $\frac{1}{4}$ are bad. Jed throws these away.
  He now has 24 oranges left. How many oranges did Jed buy?

Chapter review

- To add (or subtract) fractions, change them to **equivalent fractions** that have the same denominator. This new denominator will be the LCM of the two denominators. Then add (or subtract) the numerators but do not change the denominator.
- To add (or subtract) **mixed numbers**, add (or subtract) the whole numbers, then add (or subtract) the fractions separately.
- To multiply fractions, convert any mixed numbers to **improper fractions**, simplify if possible, then multiply the numerators and multiply the denominators.
- To divide fractions, convert any mixed numbers to improper fractions, convert divide to multiply and **invert** the second fraction, then multiply the numerators and multiply the denominators.
- You can use your knowledge of fractions to solve problems from real life.

Review exercise

1   Many wage earners work a fixed number of hours, typically 35 hours a week.
    If they are required to work more than this, they are paid overtime. For example, overtime paid at
    'time and a half' would mean someone normally earning £8 per hour would receive £12 per hour for
    the extra hours.
    Copy and complete the table for the following workers.

Name	Hourly rate	Overtime at time and a half	Overtime at double time
Aaron	£8.50		
Chi	£12.00		
Mahmood	£14.40		

D

2   Work out
    a $\frac{2}{5} + \frac{1}{4}$

    b $\frac{5}{6} + \frac{3}{4}$

    c $2\frac{1}{4} + 3\frac{1}{3}$

    d $51\frac{7}{8} + 32\frac{2}{3}$

ResultsPlus
Exam Question Report

33% of students answered this sort of question
poorly because they simply added the numerators
and added the denominators.

A03

3   a   Using the table above find Aaron's weekly wage if he works 35 hours at £8.50 per hour and
        6 hours' overtime at time and a half.
    Chi normally works 36 hours a week and is paid for any overtime at time and a quarter.
    b   One week she earned £522. How many extra hours did she work?

4   Work out   a $\frac{6}{7} \div \frac{1}{3}$ b $\frac{4}{9} \div \frac{2}{5}$ c $\frac{3}{8} \div \frac{1}{4}$ d $\frac{1}{5} \div \frac{3}{7}$

C

5   Work out   a $4\frac{1}{2} \div \frac{1}{3}$ b $3\frac{1}{5} \div \frac{2}{6}$ c $2\frac{9}{10} \div \frac{3}{4}$ d $3\frac{8}{13} \div \frac{2}{9}$

6   Work out

    a $\frac{3}{4} - \frac{2}{3}$

    b $5\frac{3}{5} - 2\frac{1}{6}$

    c $4\frac{3}{4} - 2\frac{5}{6}$

ResultsPlus
Exam Question Report

71% of students answered this sort of question
well because they dealt with the integers and
fractions separately.

7

$1\frac{2}{3}$ m   Diagram **NOT**
accurately drawn

$2\frac{1}{2}$ m

    a   Work out the area of this rectangle. [*Hint:* Area of rectangle = length × width.]
    b   Work out the perimeter of this rectangle.
    c   Work out the difference in lengths between the shortest and the longest side.

8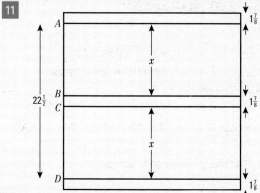

Diagram **NOT** accurately drawn

$2\frac{1}{2}$ cm $3\frac{5}{8}$ cm

The diagram represents a part of a machine.

In order to fit the machine, the part must be between $6\frac{1}{16}$ cm and $6\frac{3}{16}$ cm long.

Will the part fit the machine?

You must explain your answer.

June 2009

9 On a farm, $\frac{3}{8}$ of the land area is used to keep sheep.

Half of the rest of the land area on the farm is used to grow crops.

The land area of the farm used to grow grass is 600 000 m².

Work out the land area of the farm used to keep sheep.

June 2009

10 The distance from Granby to Hightown is $3\frac{2}{3}$ miles. The distance from Hightown to Islely is $2\frac{1}{2}$ miles.

Jim walks from Granby to Islely via Hightown. He stops for a rest when he has walked half the total distance. How far has he walked when he stops for his rest?

11

A

$1\frac{7}{8}$

x

$22\frac{1}{2}$ *B*
       C $1\frac{7}{8}$

x

D $1\frac{7}{8}$

a Here is a design for a book case with two shelves. All the measurements are in inches.

The gap *AB* is the same as the gap *CD*. Work out the value of x.

b In another design the bookcase is the same except the middle shelf has been moved so that the gap *AB* is twice the gap *CD*.

Find the size of the gap *AB*.

12

a	$\frac{1}{15}$	$\frac{2}{5}$
c	b	$\frac{7}{15}$
d	$\frac{3}{5}$	$\frac{2}{15}$

This is a magic square. The sum of the three numbers in each row, each column and each diagonal is the same.

Work out the value of a, b, c and d.

13 A scientist wants to estimate how many fish there are in a large pond.

He catches 40 one day, tags them and puts them back into the pond.

Next week he again catches 40 fish, 25 of which are tagged.

Estimate how many fish there are in the pond.

14 There are 960 pupils in a school.

$\frac{5}{8}$ of the pupils are in lower school.

$\frac{7}{12}$ of the pupils in the lower school are girls.

Work out the number of girls in the lower school.

C
A03

A02
A03

A03

A03

A03

A02
A03

B

A03

15 The diagram shows a square *ABCD*.

The points *E* and *F* are the midpoints of sides *AD* and *CD* respectively.

What fraction of the square are the triangles:

a *ABE* b *DEF* c *BCF* d *BEF*?

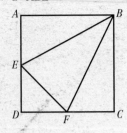

3 DECIMALS AND ESTIMATION

In athletics, sprinters often finish a race with only split seconds between them. Times are therefore given as decimals to one hundredth of a second in order to find the winner.

⊙ Objectives

In this chapter you will:
- convert between decimals and fractions
- carry out arithmetic using decimals
- write numbers correct to a given number of decimal places
- write numbers correct to a given number of significant figures.

- work out estimates to calculations by rounding
- use one calculation to find the answer to another

◁▷ Before you start

You need to be able to:
- understand and use fractions
- add and subtract decimals.

3.1 Conversion between fractions and decimals

Objectives

- You can convert between decimals and fractions.
- You can order integers, decimals and fractions.

Why do this?

Parts of a whole can be given as fractions or decimals. You need to be able to convert between these to make a comparison. One and a half kilograms can be written as 1.5 kilograms.

Get Ready

1. Write this set of numbers in order of size, smallest first.
   8.092, 8.9, 8.02, 8.09, 8.2, 8.29, 8.92
2. Write these fractions in their simplest form.
   a $\frac{15}{45}$ **b** $\frac{42}{72}$ **c** $\frac{175}{200}$

Key Points

- Decimals are used as one way of writing parts of a whole number.
- The decimal point separates the whole number part from the part that is less than 1.
- A **terminating decimal** is a decimal which ends.
  0.34, 0.276 and 5.089 are terminating decimals.
- A **recurring decimal** is one in which one or more digits repeat.
  0.111 111..., 0.563 563 563..., 8.564 444... are all recurring decimals.
- All fractions can be changed into a decimal. To work out if a fraction will be represented by a terminating or a recurring decimal:
  - write the fraction in its simplest form
  - write the denominator of the fraction in terms of its prime factors (see Section 1.1)
  - if these prime factors are only 2s and/or 5s then the fraction will convert to a terminating decimal
  - if any prime number other than 2 or 5 is a factor then the fraction will convert to a recurring decimal.
- To convert a fraction to a decimal, either divide the numerator by the denominator, or, for a terminating decimal, create an equivalent fraction in tenths or hundredths. For example, $\frac{3}{5} = 3 \div 5 = 0.6$, or $\frac{3}{5} = \frac{6}{10} = 0.6$. $\frac{1}{3} = 1 \div 3 = 0.\dot{3}$.
- Here are some common fraction-to-decimal conversions that you should remember.

Decimal	0.01	0.1	0.2	0.25	0.$\dot{3}$	0.5	0.75
Fraction	$\frac{1}{100}$	$\frac{1}{10}$	$\frac{1}{5}$	$\frac{1}{4}$	$\frac{1}{3}$	$\frac{1}{2}$	$\frac{3}{4}$

Example 1

Work out whether these fractions will convert to terminating or recurring decimals.

    a $\frac{9}{96}$ **b** $\frac{8}{30}$ **c** $3\frac{7}{20}$

a $\frac{9}{96} = \frac{3}{32}$ ⟵   Simplify the fraction.

$32 = 2 \times 2 \times 2 \times 2 \times 2$ ⟵

The only prime factor is 2.   Write the denominator in terms of its prime factors.

$\frac{9}{96}$ will convert to a terminating decimal.

b $\frac{8}{30} = \frac{4}{15}$ ← Simplify the fraction.

$15 = 3 \times 5$ ← Write the denominator in terms of its prime factors.
3 is a prime factor as well as 5.

$\frac{8}{30}$ will convert to a recurring decimal. ← As 3 is a factor of 30.

c $\frac{7}{20}$ ← Just look at the fraction part.

$20 = 2 \times 2 \times 5$ ← Write the denominator in terms of its prime factors.

$\frac{7}{20}$ will convert to a terminating decimal. ← As only 2 and 5 are the factors.

Example 2 ▶ Rearrange these numbers in order of size, smallest first.

$\frac{2}{5}, 0.34, \frac{3}{10}, 0.41$ ← Convert the fractions to decimals.

$\frac{2}{5} = \frac{4}{10} = 0.4$ ← Multiply the numerator and denominator number by 2 to create an equivalent fraction as a tenth.

$\frac{3}{10} = 0.3$ ← As the denominator is already 10, the fraction can immediately be written as a decimal.

$0.3, 0.34, 0.4, 0.41$ ← Arrange all the decimals in order of size.

$\frac{3}{10}, 0.34, \frac{2}{5}, 0.41$ ← Write the original numbers in order of size.

Exercise 3A

Questions in this chapter are targeted at the grades indicated.

1 Write each of the set of numbers in order, starting with the smallest.
$\frac{9}{10}, 0.8, 0.85, \frac{86}{100}, 0.98$

2 By writing the denominator in terms of its prime factors, decide whether these fractions will convert to recurring or terminating decimals.
a $\frac{9}{40}$ b $\frac{17}{32}$ c $\frac{8}{45}$
d $\frac{13}{42}$ e $\frac{6}{125}$ f $\frac{37}{60}$

3 Linda says that $\frac{17}{48}$ can be converted to a terminating decimal.
Mitch says that the fraction converts to a recurring decimal.
Who is correct? You must give a reason for your answer.

D

3.2 Carrying out arithmetic using decimals

Objectives

- You can multiply and divide decimals by whole numbers.
- You can multiply numbers with up to two decimal places.
- You can divide numbers with up to two decimal places.

Why do this?

To work out how much your bill is in a café, you'll often have to add prices that have decimals in them.

Get Ready

1. Work out **a** $13.1 + 5.69$ **b** $8.6 - 3.42$ **c** $37 - 9.86 + 5.6$

Key Points

- To multiply by a decimal, do the multiplication with whole numbers and then decide on the position of the decimal point.
- To divide a number by a decimal, multiply both the number and the decimal by a power of 10 (10, 100, 1000 …) to make the decimal a whole number. It is much easier to divide by a whole number than by a decimal.

Example 3   Multiply 5.12 by 4.6

$$
\begin{array}{r}
512 \\
\times\ \ 46 \\
\hline
3072 \\
20480 \\
\hline
23552
\end{array}
$$

Ignore the decimal points and do the multiplication with whole numbers.

Estimate $= 5 \times 5$
       $= 25$

An estimate for the answer is 5×5.

$5.12 \times 4.6 = 23.552$

This means that the decimal point will go between the 3 and the 5 as 23.552 is close to 25.

$5.12 \times 4.6 = 23.552$

The number of decimal places in the answer is 3, which is the same as the total number of decimal places in the question.
This rule is another way of finding the position of the decimal point in the answer.

Example 4   Divide 20 by 0.4

$$\dfrac{20}{0.4} \overset{\times 10}{\underset{\times 10}{=}} \dfrac{200}{4}$$

Convert the denominator number into a whole number.
$0.4 \times 10 = 4$
So multiply the numerator and denominator numbers in the fraction by 10.

$$
\begin{array}{r}
50 \\
4\overline{)200}
\end{array}
$$

Now divide 200 by 4.

so

$20 \div 0.4 = 50$

Example 5 Divide 4.152 by 1.2

$$\frac{4.152}{1.2} = \frac{41.52}{12}$$

×10

$$\frac{4.152}{1.2} = \frac{41.52}{12}$$

×10

To make 1.2 into a whole number, multiply it by 10.

ResultsPlus
Examiner's Tip

It does not matter that the top number is still a decimal.

$$12\overline{)41.^55^72}$$ 3. 4 6

Divide 41.52 by 12.

so 4.152 ÷ 1.2 = 3.46

Exercise 3B

1 Work out
   a 0.3 × 0.4
   b 0.006 × 0.2
   c 0.8 × 0.05
   d 0.09 × 0.07

2 Work out
   a 6.34 × 0.4
   b 4.21 × 0.3
   c 0.723 × 0.06
   d 3.15 × 0.8
   e 3.1 × 4.2
   f 0.36 × 1.4
   g 0.064 × 0.73
   h 0.095 × 3.4

3 Work out the cost of 0.6 kg of carrots at 25p per kilogram.

4 Work out the cost of 1.6 m of material at £4.29 per metre.

5 Work out
   a 12 ÷ 0.2
   b 4.2 ÷ 0.3
   c 19.2 ÷ 0.03
   d 26 ÷ 0.4
   e 5 ÷ 0.2
   f 6.12 ÷ 0.003
   g 0.035 ÷ 0.7
   h 0.008 28 ÷ 0.09

6 Work out
   a 34.65 ÷ 0.15
   b 160.5 ÷ 0.25
   c 0.8673 ÷ 0.021
   d 9.706 ÷ 0.23

7 Five people share £130.65 equally. Work out how much each person will get.

8 A bottle of lemonade holds 1.5 litres. A glass will hold 0.3 litres.
   How many glasses can be filled from the bottle of lemonade?

3.3 Rounding and decimal places

⊙ Objective

● You can write a number correct to a given number of decimal places.

❓ Why do this?

When you find a fraction of an amount of money, you may need to round the answer to two decimal places.

⬦ Get Ready

1. Work out without a calculator
   a 9.62 × 0.06
   b 9.62 ÷ 0.06
   c £9.87 ÷ 7

Key Points

● Decimals can be **rounded** to a given number of decimal places.

6.48 = 6.5 correct to 1 decimal place ←	Round up because 6.48 is closer to 6.5 than to 6.4
0.0748 = 0.07 correct to 2 decimal places ←	Round down because 0.0748 is closer to 0.07 than to 0.08
1.2475 = 1.248 correct to 3 decimal places ←	If the figure in the fourth decimal place is 5 or more then round up

Example 6 Write the following numbers correct to 2 decimal places.

   a 6.789   b 0.007   c 1.2999

a 6.79
b 0.01
c 1.30 ← | Note the difference between 1.3 (1 decimal place) and 1.30 (2 decimal places) |

Exercise 3C

1 Write the following numbers correct to 1 decimal place (1 d.p.).
   a 6.38   b 5.66   c 16.949   d 0.067   e 0.99

2 Write the following numbers correct to 2 decimal places (2 d.p.).
   a 5.667   b 8.0582   c 0.125   d 3.044   e 0.076

3 Write the following numbers correct to 3 decimal places (3 d.p.).
   a 6.4458   b 0.0792   c 5.0792   d 6.0079   e 0.0199

3.4 Significant figures

Objective

○ You can round a number correct to a given number of significant figures.

Why do this?

Sometimes you don't need to give the exact amount, you just need to give someone a round figure. For instance you might say that around 150 people were at a concert.

Get Ready

1. What is the value of the digit in the tenths column in these numbers?   a 6.38   b 4.07   c 3.99
2. Write three numbers with an '8' in the units column.

Key Points

● To write a number correct to 3 **significant figures** (3 s.f.), write down the first 3 figures, rounding up the last figure if the figure after it would be 5 or more. If necessary ignore any leading zeros.
● Leading zeros in decimals are not counted as significant.

Example 7 Round 436 to: **a** 2 significant figures
b 1 significant figure.

a 436 = 440 correct to 2 significant figures. ← Round up because 436 is closer to 440 than to 430.

b 436 = 400 correct to 1 significant figure. ← Round down because 436 is closer to 400 than 500.

Example 8 Round 0.0258 to: **a** 2 s.f. **b** 1 s.f.

a 0.0258 = 0.026 correct to 2 significant figures. ← The 8 means that the 5 will be rounded up to a 6.

b 0.0258 = 0.03 correct to 1 significant figure. ← The 5 means that the 2 will be rounded up to a 3.

Example 9 Write the following numbers correct to: **a** 3 significant figures **b** 2 significant figures.
i 2788 **ii** 4.7084 **iii** 0.006 675

a i 2790 (3 s.f.) **ii** 4.71 (3 s.f.) **iii** 0.006 68 (3 s.f.)

b i 2800 (2 s.f.) **ii** 4.7 (2 s.f.) **iii** 0.0067 (2 s.f.)

Example 10 Write the following numbers correct to 2 significant figures.
a 7995 **b** 4.996 **c** 0.000 99

a 8000 (2 s.f.)

b 5.0 (2 s.f.) ← Note the difference between 5 (1 significant figure) and 5.0 (2 significant figures).

c 0.0010 (2 s.f.)

ResultsPlus
Examiner's Tip

Although there may only be one non-zero figure in your answer, the numbers could still be correct to 2 significant figures.

Exercise 3D

1 Write the following numbers correct to 2 significant figures.
a 3867 **b** 234.7 **c** 45.53 **d** 6.48 **e** 5.079 **f** −0.4318

2 Write the following numbers correct to 3 significant figures.
a 2496 **b** 38.98 **c** 4.895 **d** 4.0899 **e** 0.010 96

3 Write the following numbers correct to 1 significant figure.
a 3499 **b** 42.62 **c** 3.008 **d** 7.92 **e** 19.8 **f** 0.982

3.5 Estimating calculations by rounding

⊙ Objective

○ You can use rounding to 1 significant figure to work out an estimate for a calculation.

? Why do this?

You can estimate the cost of buying 29 T-shirts at £5.99 using rounding.

⬆ Get Ready

1. Write correct to 1 significant figure
   - **a** 4555
   - **b** 16.8
   - **c** −6.7

2. Work out
   - **a** 20×30
   - **b** 50×300
   - **c** $600 \times 30 \times 30$

3. Work out
   - **a** $\dfrac{600}{10}$
   - **b** $\dfrac{6000}{20}$
   - **c** $\dfrac{50}{2000}$

🔍 Key Point

◉ A method of estimating the answer to a calculation is to write all numbers correct to 1 significant figure and then do the calculation.

Example 11 Work out an estimate for the area of a rectangle 38.6 cm by 12.2 cm.

Approximate area = $40 \times 10 = 400$ cm² ← | 38.6 = 40 (1 sig fig)
| 12.2 = 10 (1 sig fig)

Example 12 **a** Calculate an estimate for the value of $\dfrac{38.9 \times 19.9}{20.3}$

b Explain why your answer is an overestimate of the true answer.

ResultsPlus

Watch Out!

Remember not to round to the nearest whole number as that is a mistake.

a $\dfrac{40 \times 20}{20} = \dfrac{800}{20} = 40$

b 38.9 and 19.9 have been rounded up, so the new numerator is larger.
20.3 has been rounded down, so the denominator is smaller and dividing by a smaller denominator results in a bigger answer.

⚙ Exercise 3E

D

1 Estimate the value of the following calculations.
   - **a** 69×58
   - **b** 112×68
   - **c** 295×19
   - **d** 4897×38
   - **e** 788×109

2 Work out estimates for each of the following calculations.
   - **a** $68 \div 1.9$
   - **b** $9.9 \div 4.9$
   - **c** $58.6 \div 6.1$
   - **d** $211.8 \div 39$
   - **e** $577 \div 97.8$

3   Work out estimates for each of the following. In each case state whether your answer is an
    overestimate or an underestimate of the true answer.

    a 189×38 b $19.9 \div 5.1$ c $61.9 \div 5.92$ d $28.4 \times 1.89 \times 4.8$

4   Work out estimates for each of the following calculations.
    State whether your answer is an underestimate or an overestimate.

    a $\dfrac{48.9 \times 9.9}{11.3}$ b $\dfrac{203.8}{9.8 \times 4.9}$ c $\dfrac{999.8}{5.1 \times 5.3}$ d $\dfrac{9.55 \times 79.9}{11.8 \times 13.03}$

5   Work out an estimate for the value of 6.4×18.8^2

3.6 Estimating calculations involving decimals

⊙ Objective

⊙ You can work out an estimate for a calculation
which involves decimals by writing all numbers
correct to 1 significant figure.

⟡ Why do this?

High-powered computers can perform calculations
with numbers that have hundreds of decimal
places. We need to round these numbers in order
to work with them and estimate calculations.

⟡ Get Ready

1. Write correct to 1 significant figure
    a 0.33   b 0.0466   c 0.001 09
2. Work out
    a 0.2×30 b 0.5×300 c $0.6 \times 0.3 \times 0.3$
3. Work out
    a $\dfrac{600}{0.1}$ b $\dfrac{6000}{0.2}$ c $\dfrac{0.5}{0.02}$

🔍 Key Point

⊙ You can round figures in a calculation to a given number of significant figures to make an estimate of the answer.
This can help you to check that your answer is reasonable.

🔍 Example 13

a Work out an estimate for the value of 0.399×208.8

b Work out an estimate of the value of $\dfrac{4.89 \times 0.088}{0.0052}$

ResultsPlus

Watch Out!

Remember not to make the mistake
of rounding a decimal down to zero.

a $0.4 \times 200 = 80$

b $\dfrac{5 \times 0.09}{0.005} = \dfrac{0.45}{0.005} = \dfrac{450}{5} = 90$

> For division by a decimal, multiply the denominator and numerator
> by a number which produces a whole number in the denominator.

Exercise 3F

1 Work out estimates for the values of

    a 6.4×0.38 **b** 0.49×0.33 **c** 12.1×0.128 **d** 0.089×0.021

2 Work out estimates for the values of

    a $\dfrac{10.45}{0.49}$ **b** $\dfrac{20.8}{0.41}$ **c** $\dfrac{81.34}{0.81}$ **d** $\dfrac{0.43}{0.12}$ **e** $\dfrac{1.067}{5.49}$

3 Work out estimates for the values of the following.
State whether your answer is an overestimate or an underestimate.

    a 5.4×0.32 **b** 0.48×0.38 **c** $\dfrac{1.22}{0.19}$ **d** $\dfrac{6.02}{0.028}$

4 Work out estimates for the values of the following.
State whether your answer is an overestimate or an underestimate.

    a $\dfrac{9.8 \times 3.9}{0.14}$ **b** $\dfrac{6.8 \times 2.9}{0.11}$ **c** $\dfrac{12.1 \times 2.3}{0.83}$ **d** $\dfrac{206 \times 13.1}{0.48}$

5 $V = LWH$, $L = 0.046$, $W = 0.053$, $H = 122$. Work out an estimate for the value of V.

3.7 Manipulating decimals

◎ Objective

● You can use one calculation to find the answer to another.

⊙ Why do this?

If you know that $1.50 is worth £1, then you could use this to calculate how many cents 10p is worth.

⬆ Get Ready

1. Work out   **a** 20×3 **b** 200×3 **c** 2000×3
2. Work out   **a** $300 \div 10$ **b** $30 \div 10$ **c** $3 \div 10$

🔍 Key Point

◉ Knowing the answer to one calculation can often be used to find the answer to a second calculation.

🔎 Example 14   Given that $\dfrac{3.46 \times 25.5}{3.4} = 25.95$, find the value of each of the following.

    a $\dfrac{346 \times 25.5}{3.4}$ **b** $\dfrac{2.595 \times 0.34}{25.5}$

a **Method 1**

$\dfrac{346 \times 25.5}{3.4} = \dfrac{3.46 \times 100 \times 25.5}{3.4}$

$\phantom{\dfrac{346 \times 25.5}{3.4}} = \dfrac{3.46 \times 25.5}{3.4} \times 100$

$\phantom{\dfrac{346 \times 25.5}{3.4}} = 25.95 \times 100$

$\phantom{\dfrac{346 \times 25.5}{3.4}} = 2595$

> The final answer will be 100 times that of the given calculation as one of the numbers on the top of the fraction is 100 times the corresponding number in the original fraction.

Method 2

$$\frac{346 \times 25.5}{3.4} = \frac{300 \times 30}{3}$$

$$= \frac{9000}{3}$$

$$= 3000$$

$$\frac{346 \times 25.5}{3.4} = 2595$$

> Write down an approximation to the given sum. The answer is approximately 3000. The number closest to this gained by moving the decimal point in the answer to the given calculation is 2595 (rather than 259.5 or 25 950 etc.).

ResultsPlus
Examiner's Tip

Round each number to 1 significant figure so you can calculate an estimate quickly and easily.

b Method 1

$$\frac{2.595 \times 0.34}{25.5} = \frac{25.95 \div 10 \times 3.4 \div 10}{25.5}$$

$$= \frac{25.95 \times 3.4}{25.5} \div 100$$

$$= 3.46 \div 100$$

$$= 0.0346$$

> Two numbers on the top of the fraction have been divided by 10 so divide the answer to the rearranged calculation by 100.

Method 2

$$\frac{2.595 \times 0.34}{25.5} = \frac{3 \times 0.3}{30}$$

$$= \frac{0.9}{30}$$

$$= 0.03$$

$$\frac{2.595 \times 0.34}{25.5} = 0.0346$$

> From the original *rearranged* calculation, the number closest to 0.03 that can be obtained by moving the decimal point in 3.46 is 0.0346.

Exercise 3G

1 Given that $6.4 \times 2.8 = 17.92$ work out

a 64×28　　　b 640×2.8　　　c 0.64×28　　　d 0.64×0.028

2 Given that $18.3 \div 1.25 = 14.64$ work out

a $183 \div 1.25$　　　b $1.83 \div 1.25$　　　c $0.183 \div 1.25$　　　d $0.183 \div 12.5$

3 Given that $\dfrac{23.2 \times 5.1}{3.4} = 34.8$ work out

a $\dfrac{23.2 \times 51}{3.4}$　　　b $\dfrac{232 \times 51}{3.4}$　　　c $\dfrac{23.2 \times 51}{34}$　　　d $\dfrac{232 \times 51}{34}$

4 Given that $23 \times 56 = 1288$ work out

a 0.23×560　　　b $1288 \div 5.6$　　　c $12.88 \div 0.23$　　　d $1288 \div (23 \times 28)$

5 Given that $884 \div 34 = 26$ work out

a $8.84 \div 340$　　　b $884 \div 2.6$　　　c $8.84 \div 260$　　　d $884 \div (3.4 \times 2.6)$

6 Given that $\dfrac{1872}{1.2^2} = 1300$ work out

a $\dfrac{1872}{12^2}$　　　b $\dfrac{18.72}{1.2^2}$　　　c $\dfrac{187.2}{0.12^2}$　　　d $\dfrac{936}{120^2}$

D

Chapter review

- Decimals are used as one way of writing parts of a whole number.
- The decimal point separates the whole number part from the part that is less than 1.
- A **terminating decimal** is a decimal which ends.
  0.34, 0.276 and 5.089 are terminating decimals.
- A **recurring decimal** is one in which one or more figures repeat.
  0.111 111..., 0.563 563 563..., 8.564 444... are all recurring decimals.
- To work out if a fraction will be represented by a terminating or a recurring decimal:
  - write the fraction in its simplest form
  - write the denominator of the fraction in terms of its prime factors
  - if these prime factors are only 2s and/or 5s then the fraction will convert to a terminating decimal
  - if any prime number other than 2 or 5 is a factor then the fraction will convert to a recurring decimal.
- To convert a fraction to a decimal, either divide the numerator by the denominator or, for a terminating decimal, create an equivalent fraction in tenths or hundredths.
- When adding or subtracting decimals, line up the decimal points first.
- To multiply by a decimal, do the multiplication with whole numbers and then decide on the position of the decimal point.
- To divide a number by a decimal, multiply both the number and the decimal by a power of 10 (10, 100, 1000 ...) to make the decimal a whole number. It is much easier to divide by a whole number than by a decimal.
- Decimals can be **rounded** to a given number of decimal places.
- To write a number correct to 3 **significant figures** (3 s.f.), write down the first 3 figures, rounding up the last figure if the figure after it would be 5 or more. If necessary ignore any leading zeros.
- Leading zeros in decimals are not counted as significant.
- A method of estimating the answer to a calculation is to write all numbers correct to 1 significant figure.
- You can round figures in a calculation to a given number of significant figures to make an estimate of the answer. This can help you to check that your answer is reasonable.
- Knowing the answer to one calculation can often be used to find the answer to a second calculation.

Decimal	Fraction
0.01	$\frac{1}{100}$
0.1	$\frac{1}{10}$
0.2	$\frac{1}{5}$
0.25	$\frac{1}{4}$
0.5	$\frac{1}{2}$
0.75	$\frac{3}{4}$

Review exercise

1 Here are the rates of pay in a company.

Grade	Basic pay for an hour's work	Overtime pay for an hour's work
Operative	£5.40	£8.10
Technician	£7.50	£11.25
Supervisor	£9.00	£13.50
Driver	£7.20	£10.80

Kaysha has a part-time job as an operative.
Last week Kaysha earned basic pay for 24 hours and overtime pay for 3 hours.
Work out Kaysha's total pay for last week.

June 2008, adapted

2 Write down which of the following, when written as decimals, are recurring.

$\frac{3}{4}$ $\frac{2}{3}$ $\frac{7}{8}$ $\frac{9}{24}$ $\frac{3}{5}$

3 Put these numbers in order. Start with the smallest number.

$\frac{3}{5}$ 0.47 $\frac{12}{25}$ $\frac{31}{50}$

4 Ethan has a '5p off per litre' voucher for use at a local petrol station.
He fills up his tank with 43 litres of petrol normally costing 104.9p per litre.
How much does he pay?

5 Write correct to 3 significant figures
  a 4778   b 106.74   c 3.228×10^{15} d 6996   e 56.97

6 Write correct to 2 significant figures
  a 45.87   b 30.72   c 0.0457   d 19.97   e 4.098

7 Write correct to 1 significant figure
  a 363   b 40.22   c 9.9×10^{17} d 0.005 48   e -3.056

8 Write the following numbers correct to 1 decimal place
  a 3.142   b 0.567   c 2.091   d 3.99

9 Use the information that
$$\frac{63 \times 99}{18^2} = \frac{77}{4}$$
Write down the value, as fractions or integers, of

  a $\dfrac{6.3 \times 9.9}{18^2}$

  b $\dfrac{6.3 \times 990}{18^2}$

  c $\dfrac{63 \times 99}{0.18^2}$

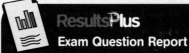

ResultsPlus
Exam Question Report

97% of students answered this sort of question well because they used the information provided.

10 Jim and Jenni are collecting money for charity. Jim collects a kilometre of 2p pieces. Jenni wants to collect 1p pieces. How many 1p pieces does she need to collect to get the same amount of money as Jim? (Diameter of 2p piece = 2.6 cm, diameter of 1p piece = 2 cm).

*11 An electricity bill includes a standard charge and an amount depending on how much you use. Avery Energy charges 12.61p per unit for the first 400 units and 15.02p per unit for any units used above this amount.

AVERY ENERGY	Total units = 573		
400 units	@ 12.61p per unit	=	£50.44
173 units	@ 15.02p per unit	=	£25.98
Standing charge, 90 days	@ 14.01p per day	=	£12.61
	Total due	=	£89.03

  a Check the amounts on the bill are correct.

D

Three electricity companies use different pricing structures.

AVERY ENERGY	
Standing charge	14.01p per day
First 400 units	@ 12.61p per unit
Additional units	@ 15.02p per unit

BRAWN POWER	
Standing charge	16.30p per day
All units charged	@ 13.60p per unit

CC ELECTRICS	
No standing charges	
First 100 units	@ 5.03p per unit
Additional units	@ 23.72p per unit

b   Advise the following people which electricity company they should use. (Take 90 days as a quarter.)
i   John is rarely in. He uses about 350 units a quarter.
ii   Vijay works from home and uses 1261 units a quarter.

12   By writing each number correct to 1 significant figure, work out an estimate for the value of
   a 49×59 b 79×51 c 4.1×5.9 d 499×691 e $6.1 \times 19.9 \times 2.8$

13   By writing each number correct to 1 significant figure, work out an estimate for the value of
   a $199 \div 39$ b $19.9 \div 4.1$ c $411 \div 4.9$ d $4991 \div 21.8$ e $19.98 \div 20.8$

14   The table gives information about the length of time in minutes and seconds, the tracks on a CD last.

Track number	1	2	3	4	5	6
Time	7 m 56 s	3 m 5 s	4 m 8 s	5 m 58 s	3 m 57 s	11 m 48 s

Work out an estimate for the total length of time for all the tracks on the CD.
Give your answer in minutes.

A02   **15**   Estimate in £s the cost of:
   a   an ice cream in Italy costing 3 euros
   b   a T-shirt in the USA costing $30
   c   a meal in Japan costing 1500 yen
   d   a drink in Turkey costing 10 lira
   e   a souvenir in Thailand costing 300 baht
   f   a camera in China costing 2000 yuan.

£1 buys you
1.1 euro
US$1.63
148 Japanese yen
2.45 Turkish lira
54.7 Thai baht
11.15 Chinese yuan

C   **A02**   **A03**   **16**   Rob's tariff for his mobile phone is shown in the box on the right.

   a   Calculate his monthly bill if he made 100 minutes of calls and 60 texts.
   b   In one particular month, the number of texts and calls were
      the same.
      If his bill was £8, how many texts did he send?

No monthly fee
Calls
15p per minute anytime
Texts
10p per text to any network

17   Work out an estimate for the value of each of these. In each case state whether your answer is an
overestimate or an underestimate.
   a $\dfrac{5.4 \times 3.2}{0.187}$ b $\dfrac{0.32}{0.00195}$ c $\dfrac{0.88 \times 0.37}{0.131}$ d $\dfrac{59 \times 36}{0.415}$ e $\dfrac{0.32 \times 320}{0.195 \times 0.012}$

18 The height of a room is 285 cm. The width of the room is 790 cm and the length is 880 cm.

A number of people are to work in the room. Building regulations state that for every person working in the room there must be 4.25 m³.

Calculate an estimate for the number of people who could work in the room.

19 Jason hired a van.

The company charges £90 per day plus the cost of the fuel used.

The van can travel 6 miles for each litre of fuel used.

Fuel costs 98.9 p for 1 litre.

On Monday, Jason hired the van and drove from London to Cardiff.

London			
155	Cardiff		
212	245	York	
413	400	193	Edinburgh

On Tuesday, Jason drove from Cardiff to Edinburgh.

On Wednesday, Jason drove from Edinburgh back to London and returned the van.

Jason used this table for information about distances between cities.

Jason thought the total cost would be about £400.

Work out the total cost of hiring the van and the fuel used.

May 2009

4 PERCENTAGES

One of the most common percentages used to compare schools is the number of students getting five GCSEs grade A*–C. Percentages can also be used to compare results across the sexes; traditionally, girls have outdone boys but the gap between the sexes looked as if it was beginning to close in 2009, with 70.5% of girls getting five GCSEs grade A*–C, compared to 63.6% of boys. This is the smallest gap since 1991.

◉ Objectives

In this chapter you will:
- find a percentage of a quantity
- find quantities after a percentage increase or decrease.

◈ Before you start

You should know:
- how to find a fraction of a quantity
- how to convert between fractions and decimals
- that percent means 'out of 100'
- how to write a percentage as a fraction or a decimal.

4.1 Working out a percentage of a quantity

◉ Objectives

- ◉ You can convert between fractions, decimals and percentages.
- ◉ You can find a percentage of a quantity.

⍰ Why do this?

Percentages are a part of our everyday language. Banks pay interest as a percentage of the money in your bank account. Tax is paid as a percentage of money earned.

⬀ Get Ready

1. Write these percentages as **i** fractions in their simplest form   **ii** decimals.
   a 20%   **b** $12\frac{1}{2}$%   **c** 60%   **d** $17\frac{1}{2}$%

◕ Key Points

- ◉ There are a number of different methods that can be used to work out a percentage of an amount.
- ◉ When not using a calculator, first work out either 10% or 1% of the amount then build up the percentage.

Example 1   Colin invests £1800.
The **interest** rate is 5% per year.
How much interest will Colin receive after 1 year?

ResultsPlus
Watch Out!

As the answer is an amount of money, remember to give two decimal places in the answer.

Method 1

$$5\% = \frac{5}{100}$$ ← Convert the percentage to a fraction.

$$\frac{5}{100} \times 1800 = 90$$ ← Multiply the amount by the fraction.

Interest will be £90

Method 2

$$5\% = 0.05$$ ← Convert the percentage to a decimal.

$$0.05 \times 1800 = 90$$ ← Multiply the amount by the decimal.

Interest will be £90

Method 3

$$1\% \text{ of } 1800 = 1800 \div 100$$
$$= 18$$ ← Work out 1% of the amount. As $1\% = \frac{1}{100}$, divide by 100.

$$5\% \text{ of } 1800 = 18 \times 5$$
$$= 90$$ ← $5\% = 5 \times 1\%$ so multiply 18 by 5.

Interest will be £90

Exercise 4A

Questions in this chapter are targeted at the grades indicated.

1 Work out
   a 30% of £600   b 15% of 40 kg   c 5% of £32.40   d 20% of 62 kg
   e 20% of £30   f 30% of 150   g 5% of 60 km   h $17\frac{1}{2}$% of £300

2 There are 150 shop assistants in a large store. 40% of the shop assistants are male.
   How many of the shop assistants are male?

3 Danya invests £250. The interest rate is 4% per year.
   How much interest will she receive after 1 year?

4 A shop has 4600 DVDs. 20% of the DVDs are thrillers.
   How many of the DVDs in the shop are thrillers?

5 There are 154 students in Year 11. 84 of these students are girls.
   50% of the girls and 10% of the boys attend Spanish lessons.
   What fraction of these Year 11 students attend Spanish lessons? Give your fraction in its simplest form.

6 Here are the rules that Mikhail uses to work out how much tax he must pay.
   Earn the first £8000 tax free.
   Pay tax of 20% on the next £35 000.
   Pay tax of 40% on the rest.
   Mikhail earns £70 000 per year. Work out how much tax Mikhail has to pay each year.

7 Jed is deciding whether to live in country A or country B. He looks at their tax rules to help him decide
   where to live.

Country A	Country B
Earn the first £20 000 tax free	Earn the first £15 000 tax free
Pay tax of 40% on the rest	Pay tax of 25% on the next £40 000
	Pay tax of 50% on the rest

   Jed expects to earn £90 000 per year. In which country will he pay less tax and by how much?

4.2 Finding the new amount after a percentage increase or decrease

Objective

● You can find quantities after a percentage increase or decrease.

Why do this?

Percentages are often used when a shop has a sale or is offering a discount.

Get Ready

1. Work out 25% of £300.
2. Work out 20% of £80.
3. Write 5 as a fraction of 20.

Key Points

● There are two methods that can be used to increase an amount by a percentage.
  ◉ You can find the percentage of that number and then add this to the starting number.
  ◉ You can use a **multiplier**.

● There are two methods that can be used to decrease an amount by a percentage.
  ◉ You can find the percentage of that number and then subtract this from the starting number.
  ◉ You can use a multiplier.

Example 2   Hugh's salary is £25 000 a year.
His salary is increased by 4%.
Work out his new salary.

Method 1

4% of £25 000 $= \frac{4}{100} \times 25\,000$

$= 1000$ ← The increase in his salary is £1000.

$25\,000 + 1000 = 26\,000$ ← Add the increase to his original salary.

Hugh's new salary is £26 000.

Method 2

$100\% + 4\% = 104\%$ ← His new salary is 104% of £25 000.

$104\% = \frac{104}{100} = 1.04$ ← 1.04 is the multiplier.

$1.04 \times 25\,000 = 26\,000$ ← Multiply 25 000 by 1.04.
(This increases 25 000 by 4%.)

Hugh's new salary is £26 000.

Example 3   a Write down the single number that you can multiply by to increase an amount by 10%.
b Increase £13 by 10%.

a $100\% + 10\% = 110\%$

$110\% = \frac{110}{100} = 1.1$

b $13 \times 1.1 = £14.30$ ← Use the multiplier worked out in part a.

Exercise 4B

1 a Increase £300 by 20%   b Increase 90 kg by 30%
   c Increase 40 km by 25%   d Increase £1200 by 15%

D

D

2   Kylie earns £210 each week. She receives a pay increase of 5%. How much does she now earn each week?

3   Mrs Barlow buys a house for £320 000. The price of the house increases by 15%. How much is the house now worth?

4   Hamid buys a television set for £640 plus VAT at $17\frac{1}{2}$%. Work out the total price including VAT.

5   Larry gets a 7% increase in pay. He was earning £21 000 a year. What is his new salary?

6   There are 1200 pupils in a school. Next year it is planned that there will be an increase of 8% in the number of pupils. How many pupils should there be at the school next year?

A02 A03

7   Jamil earns £1000 per month. He gets a pay rise of 4%.
Sam earns £300 per week. He gets a pay rise of 1.5%.
Who now earns the biggest salary?

A02 A03

8   Sarah and Jack both invest £4000 for one year. Sarah puts all her money in a building society which pays 5% interest. Jack invests half of his money in a bank that pays 4% interest. He buys a premium bond with the rest of his money. The premium bond does not earn any interest but Jack wins a £150 prize. Who has the most money at the end of the year?

🔍 **Example 4**   The value of a car depreciates by 15% each year.
The value of a car when new is £14 000.
Work out the value of the car after 1 year.

Depreciates means that the value of the car decreases.

Method 1

15% of £14 000 = $\frac{15}{100}$ × 14 000   ← The depreciation in 1 year is £2100.

= 2100

14 000 − 2100 = 11 900   ← Subtract to work out the new value.

Value after 1 year = £11 900.

Method 2

100% − 15% = 85%   ← The final value is 85% of the original value.

85% = $\frac{85}{100}$ = 0.85   ← 0.85 is the multiplier.

0.85 × 14 000 = 11 900   ← Multiply the original amount by 0.85.

Value after 1 year = £11 900.

Exercise 4C

1 a Decrease £400 by 10%   b Decrease 200 kg by 15%

   c Decrease 70 m by 30%   d Decrease £1500 by 45%

2 Jeevan buys a van for £20 000. At the end of 1 year the value of the van has depreciated by 15%. How much is the van now worth?

3 Last year a total of 1650 people came to see a school play. This year, attendance was down by 10%. How many people came to see the school play this year?

4 In a sale prices are reduced by 30%. The normal price of a washing machine is £650. How much would it cost in the sale?

5 Kaz books a holiday. Last year the price of the same holiday was £840. This year prices are 15% less than last year. How much will Kaz pay for the same holiday this year?

6 Brian weighs 94 kg before going on a diet. He sets himself a target of losing 5% of his original weight. What is his target weight?

Chapter review

- There are a number of different methods that can be used to work out a percentage of an amount.
- When not using a calculator, first work out either 10% or 1% of the amount and build up the percentage.
- There are two methods that can be used to increase (or decrease) an amount by a percentage.
  - You can find the percentage of that number and then add this to (or subtract this from) the starting number.
  - You can use a **multiplier**.

Review exercise

1 There are 800 students at Prestfield School.
45% of these 800 students are girls.
   a Work out 45% of 800.
There are 176 students in Year 10.
   b Write 176 out of 800 as a percentage.   *June 2004*

2 A car tyre costs £80 plus VAT at 17.5%.
Work out the total cost of the tyre.   *June 2008*

3 Work out: a 30% of £800   b 25% of 20 kg   c 15% of £70   d $17\frac{1}{2}$% of £60.

4 The normal price of a cat basket is £40.
In a sale, the price of the cat basket is reduced by 15%.
Work out the sale price of the cat basket.

D

5 This item appeared in a newspaper.

COWS PRODUCE 3% MORE MILK
A farmer has found that when his cow listened to classical music the milk it produced increased by 3%.

The cow's usual milk production rate was 18 litres per day. Calculate the amount of milk produced by the cow when it listened to classical music.

6 A farmer has a rectangular field. He makes the field 20% longer.
If he wants to keep the same area, what would he have to reduce the width by?

* 7 The same barbeque set is sold in three different shops.
Here are the price labels shown on each barbeque set.

Shop A	Shop B	Shop C
£680.00 (inc. VAT)	£640.00 (inc. VAT)	£435.00
Get $\frac{1}{4}$ off when you buy this barbeque set	Now with 20% discount	Plus 17.5% VAT

Which barbeque is the best buy?

* 8 Barry has been asked to compare the pay for four similar jobs advertised in a newspaper.

Able Computer Sales	Beta IT Support
Sales Assistant	Sales Consultant
You will spend time in the field, working both from our Manchester headquarters and from home in the North West region.	Full time: 30 hours per week
	Pay: £15 per hour
	Tele-sales based in our new offices.
Pay: £23 000 per annum	Daily hours variable.
Compu Systems	Digital Hardware
Sales Agent	Sales Adviser
As a sales agent your pay will be £1800 per month, plus commission of 1% of monthly sales. You can expect to make monthly sales to a minimum value of £22 000.	You will be part of a team with a salary of £20 000 per annum + team bonus. Team bonus last year was 20% of salary.

Which job pays the most?

9 Jessica's annual income is £12 000.
She pays $\frac{1}{4}$ of the £12 000 in rent.
She spends 10% of the £12 000 on clothes.
Work out how much of the £12 000 Jessica has left.

10 Bytes is a shop that sells computers.
In 2003, Bytes sold 620 computers.
In 2004, Bytes sold 708 computers.
Work out the percentage increase in the number of computers sold.
Give your answer correct to 1 decimal place.

Nov 2005

5 INDICES, STANDARD FORM AND SURDS

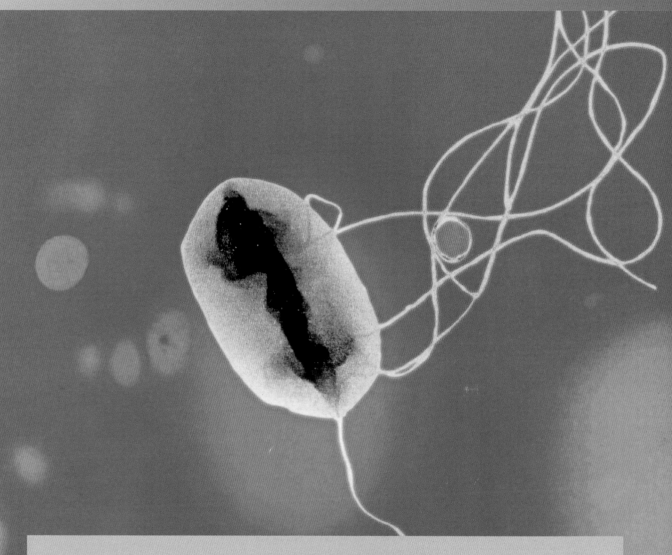

The photo shows a male *Escheria coli* bacteria. You may have heard of e-coli. These bacteria are commonly known in relation to food poisoning as they can cause serious illness. Each bacterium is about a millionth of a metre, or 0.000001 m long. Standard form allows us to write both very large and very small numbers in a more useful form.

Objectives

In this chapter you will:

- work out the value of an expression with zero, negative or fractional indices
- convert between standard form and ordinary numbers
- calculate with numbers in standard form
- manipulate surds.

Before you start

You need to be able to:

- use the index laws.

5.1 Using zero and negative powers

Objectives

- You know that $n^0 = 1$ when $n \neq 0$.
- You know the meaning of negative indices.

Why do this?

If you are x metres from a live band, the volume of sound they are producing is directly proportional to x^{-2}. This means that if you halve your distance from the band, the music will get four times as loud.

Get Ready

Work out   **1.** $\frac{1}{4} \times \frac{1}{4}$ **2.** $\frac{3}{5} \times \frac{3}{5}$ **3.** -2^3

Key Points

- For non-zero values of a
  $$a^0 = 1$$
- For any number n
  $$a^{-n} = \frac{1}{a^n}$$

Example 1

Work out the value of   **a** 3^0 **b** 5^{-1} **c** 6^{-2} **d** $\left(\frac{2}{5}\right)^{-2}$

a $3^0 = 1$ ← Any number to the power of zero is 1.

b $5^{-1} = \frac{1}{5}$ ← Use the rule $a^{-n} = \frac{1}{a^n}$

c $6^{-2} = \frac{1}{6^2}$
$\phantom{6^{-2}} = \frac{1}{36}$ ← $6^2 = 6 \times 6 = 36$

d $\left(\frac{2}{5}\right)^{-2} = \frac{1}{\left(\frac{2}{5}\right)^2}$ ← To work out the reciprocal of a fraction, turn the fraction upside down. Square the number on the top and the number on the bottom of the fraction.
$\phantom{\left(\frac{2}{5}\right)^{-2}} = \left(\frac{5}{2}\right)^2$
$\phantom{\left(\frac{2}{5}\right)^{-2}} = \frac{25}{4}$

ResultsPlus
Examiner's Tip

Do not convert the fraction to a decimal. It is much easier to square the numbers in a fraction than it is to square a decimal.

Exercise 5A

Questions in this chapter are targeted at the grades indicated.

1 Write down the value of these expressions.

    a 7^0 **b** 8^{-1} **c** 5^{-1} **d** 4^0

    e $(-2)^{-3}$ **f** 9^{-2} **g** 10^{-4} **h** 145^0

    i $(-3)^{-2}$ **j** $(-8)^0$ **k** 16^0 **l** 10^{-6}

B

2 Work out the value of these expressions.

a $\left(\frac{1}{3}\right)^{-1}$ b $\left(\frac{2}{7}\right)^{-1}$ c $\left(\frac{1}{7}\right)^{-2}$ d $\left(\frac{1}{4}\right)^{-3}$

e $(0.25)^{-2}$ f $\left(\frac{2}{5}\right)^{-3}$ g $\left(\frac{5}{3}\right)^{0}$ h $\left(\frac{9}{5}\right)^{-1}$

i $\left(1\frac{2}{5}\right)^{-2}$ j $\left(1\frac{1}{3}\right)^{-3}$ k $(0.1)^{-4}$ l $(0.2)^{-3}$

5.2 Using standard form

◎ Objectives

○ You can convert between ordinary numbers and standard form.
○ You can calculate with numbers in standard form.

◈ Why do this?

Astronomers use standard form to record large measurements. The Sun's diameter is about 1.392×10^6 km. Biologists working with micro-organisms sometimes use standard form to record their very small sizes, like 2.1×10^{-4} cm.

◈ Get Ready

1. Work out   a 10^3 b 10^{-2}
2. Write 10 000 as a power of 10.
3. Work out $2.35 \times 10\,000$.

◉ Key Points

◉ **Standard form** is used to represent very large (or very small) numbers.
◉ A number is in standard form when it is in the form $a \times 10^n$ where $1 \leqslant a < 10$ and n is an integer.
◉ A number in standard form looks like this.

$$6.7 \times 10^4$$

This part is written as a number between 1 and 10.

This part is written as a power of 10.

◉ These numbers are all in standard form: 4.5×10^2, 9×10^{-8}, 1.2657×10^6.
◉ These numbers are not in standard form because the first number is not between 1 and 10: 67×10^9, 0.087×10^3.
◉ It is often easier to multiply and divide very large or very small numbers, or estimate a calculation, if the numbers are written in standard form.

Example 2   Write these numbers in standard form.   a 50 000   b 34 600 000   c 682.5

a $50\,000 = 5 \times 10\,000$
$= 5 \times 10^4$

b $34\,600\,000 = 3.46 \times 10\,000\,000$
$= 3.46 \times 10^7$

> Use 3.46 not 34.6 or 346 as 3.46 is between 1 and 10.

c $682.5 = 6.825 \times 100$
$= 6.825 \times 10^2$

Example 3 Write as an ordinary number. a 8.1×10^5 b 6×10^8

a $8.1 \times 10^5 = 8.1 \times 100\,000$
$\qquad = 810\,000$
b $6 \times 10^8 = 6 \times 100\,000\,000$
$\qquad = 600\,000\,000$

Exercise 5B

1 Write these numbers in standard form.
   a 700 000   b 600   c 2000   d 900 000 000   e 80 000

2 Write these as ordinary numbers.
   a 6×10^5 b 1×10^4 c 8×10^5 d 3×10^8 e 7×10^1

3 Write these numbers in standard form.
   a 43 000   b 561 000   c 56   d 34.7   e 60

4 Write these as ordinary numbers.
   a 3.96×10^4 b 6.8×10^7 c 8.02×10^3 d 5.7×10^1 e 9.23×10^0

5 In 2008 there were approximately 7 000 000 000 people in the world. Write this number in standard form.

6 The circumference of Earth is approximately 40 000 km. Write this number in standard form.

Example 4 Write these in standard form.
a 0.000 000 006   b 0.000 56

a $0.000\,000\,006 = 6 \times 0.000\,000\,001$ ← $0.000\,000\,001$ is equivalent to $\frac{1}{1\,000\,000\,000}$.
$\qquad = 6 \times \frac{1}{1\,000\,000\,000}$
$\qquad = 6 \times \frac{1}{10^9}$ ← Using $a^{-n} = \frac{1}{a^n}$
$\qquad = 6 \times 10^{-9}$

b $0.000\,56 = 5.6 \times 0.0001$
$\qquad = 5.6 \times \frac{1}{10\,000}$
$\qquad = 5.6 \times \frac{1}{10^4}$ ← Use 5.6 rather than 56 as 5.6 is between 1 and 10.
$\qquad = 5.6 \times 10^{-4}$

Example 5 Write these as ordinary numbers.
a 3×10^{-6} b 1.5×10^{-3}

a $3 \times 10^{-6} = \frac{3}{10^6}$
$\qquad = \frac{3}{1\,000\,000}$
$\qquad = 0.000\,003$

b $1.5 \times 10^{-3} = \frac{1.5}{10^3}$
$\qquad = \frac{15}{10\,000}$
$\qquad = 0.0015$

Exercise 5C

1   Write these numbers in standard form.
    a 0.005   b 0.04   c 0.000 007   d 0.9   e 0.0008

2   Write these as ordinary numbers.
    a 6×10^{-5} b 8×10^{-2} c 5×10^{-7} d 3×10^{-1} e 1×10^{-8}

3   Write these numbers in standard form.
    a 0.0047   b 0.987   c 0.000 803 4   d 0.000 15   e 0.601

4   Write these as ordinary numbers.
    a 8.43×10^{-5} b 2.01×10^{-2} c 4.2×10^{-7} d 7.854×10^{-1} e 9.4×10^{-4}

5   Write these numbers in standard form.
    a 457 000   b 0.0023   c 0.0003   d 2 356 000   e 0.782
    f 89 000   g 200   h 0.005 26   i 6034   j 0.000 008 73

6   Write these as ordinary numbers.
    a 4.12×10^{-4} b 3×10^{3} c 2.065×10^{7} d 4×10^{-6} e 3.27×10^{8}
    f 7.5×10^{-1} g 1.5623×10^{2} h 5.12×10^{-7} i 2.7×10^{5} j 6.12×10^{-1}

7   1 micron is 0.000 001 of a metre. Write down the size of a micron, in metres, in standard form.

8   A particle of sand has a diameter of 0.0625 mm. Write this number in standard form.

Example 6   Write in standard form.
            a 40×10^{2} b 0.008×10^{-2}

Method 1

a $40 \times 10^{2} = 4 \times 10^{1} \times 10^{2}$ ← Write 40 in standard form.
   $\qquad\qquad = 4 \times 10^{1+2}$ Use the rule $a^{m} \times a^{n} = a^{m+n}$.
   $\qquad\qquad = 4 \times 10^{3}$

b $0.008 \times 10^{-2} = 8 \times 10^{-3} \times 10^{-2}$ ← Write 0.008 in standard form.
   $\qquad\qquad\quad = 8 \times 10^{-3 + -2}$ Use $a^{m} \times a^{n} = a^{m+n}$.
   $\qquad\qquad\quad = 8 \times 10^{-5}$

ResultsPlus
Examiner's Tip

The power of 10 tells you how many 0s there are.
$10^{2} = 100$ 2 zeros
$10^{-2} = 0.01$ 2 zeros

Method 2

a $40 \times 10^{2} = 40 \times 100$
   $\qquad\qquad = 4000$
   $\qquad\qquad = 4 \times 10^{3}$ ← Work out the calculation.
                                        Convert the answer into standard form.

b $0.008 \times 10^{-2} = 0.008 \times \frac{1}{100}$
   $\qquad\qquad\quad = 0.008 \div 100$
   $\qquad\qquad\quad = 0.000 08$ ← Use the rule $a^{-n} = \frac{1}{a^{n}}$.
   $\qquad\qquad\quad = 8 \times 10^{-5}$ Multiplying by $\frac{1}{100}$ is the same as dividing by 100.

Exercise 5D

B

1   Write these in standard form.
    a 45×10^3 b 980×10^{-3} c 3400×10^{-2} d 186×10^{10}

2   Write these in standard form.
    a 0.009×10^5 b 0.045×10^6 c 0.3708×10^{-12} d 0.006×10^{-7}

3   Which of these numbers are in standard form?
    If a number is not in standard form, rewrite it so that it is.
    a 7.8×10^4 b 890×10^6 c 13.2×10^{-5}
    d 0.56×10^9 e $60\,000 \times 10^{-8}$ f 8.901×10^{-7}
    g $0.040\,05 \times 10^{-10}$ h 9080×10^{15} i 6.002×10^5
    j 0.0046×10^8 k $67\,000 \times 10^{-3}$ l 0.004×10^3

4   Write these numbers in order of size. Start with the smallest number.
    $6.3 \times 10^6, 0.637 \times 10^7, 6\,290\,000, 63.4 \times 10^5$

5   Write these numbers in order of size. Start with the smallest number.
    $0.034 \times 10^{-2}, 3.35 \times 10^{-5}, 0.000\,033, 37 \times 10^{-4}$

Example 7 ▶   Work out $(3 \times 10^6) \times (4 \times 10^3)$ giving your answer in standard form.

$$(3 \times 10^6) \times (4 \times 10^3) = 3 \times 4 \times 10^6 \times 10^3$$
$$= 12 \times 10^9$$
$$= 1.2 \times 10^1 \times 10^9$$
$$= 1.2 \times 10^{10}$$

> Rearrange the expression so the powers of 10 are together.
> Multiply the numbers.
> Use $a^m \times a^n = a^{m+n}$ to multiply the powers of 10.
> 12×10^9 is not in standard form.
> Write your final answer in standard form.

Exercise 5E

A

1   Work out and give your answer in standard form.
    a $(4 \times 10^8) \times (2 \times 10^3)$ b $(6 \times 10^5) \times (1.5 \times 10^3)$
    c $(4 \times 10^{-7}) \times (3 \times 10^5)$ d $(8.6 \times 10^8) \div (2 \times 10^{13})$
    e $(1 \times 10^{12}) \div (4 \times 10^3)$ f $(7 \times 10^{-9}) \div (7 \times 10^{-5})$

2   Write these in standard form.
    a $(2 \times 10^5)^2$ b $(5 \times 10^{-5})^2$ c $(4 \times 10^6)^2$ d $(7 \times 10^{-8})^2$

3   The base of a microchip is in the shape of a rectangle. Its length is 2×10^3 mm and its width is
    1.6×10^{-3} mm. Find the area of the base. Give your answer in mm^2 in standard form.

5.3 Working with fractional indices

⊙ Objective

⊙ You know the meaning of fractional indices.

⊘ Why do this?

Fractional indices are used when you model the rates at which things vibrate, such as your voice box.

⬆ Get Ready

Work out   **1.** $\sqrt[3]{1000}$ **2.** $\sqrt[3]{8}$ **3.** $\sqrt[3]{-27}$

🌐 Key Points

⊙ Indices can be fractions. In general,
$$a^{\frac{1}{n}} = \sqrt[n]{a}$$

⊙ In particular, this means that
$$a^{\frac{1}{2}} = \sqrt{a} \text{ and } a^{\frac{1}{3}} = \sqrt[3]{a}$$

Example 8

Find the value of the following

a $25^{\frac{1}{2}}$ b $(-1000)^{\frac{1}{3}}$ c $16^{-0.25}$

a $25^{\frac{1}{2}} = \sqrt{25}$ ← The square root of 25 is 5 because $5 \times 5 = 25$.
$= 5$

b $(-1000)^{\frac{1}{3}} = \sqrt[3]{-1000}$ ← The cube root of -1000 is -10 because $-10 \times -10 \times -10 = -1000$.
$= -10$

c $16^{-0.25} = 16^{-\frac{1}{4}}$ ← Convert the decimal into a fraction $0.25 = \frac{1}{4}$. Use the rule $a^{-n} = \frac{1}{a^n}$.

$= \dfrac{1}{16^{\frac{1}{4}}}$

$= \dfrac{1}{\sqrt[4]{16}}$ ← $16^{\frac{1}{4}} = \sqrt[4]{16} = 2$ because $2^4 = 16$

$= \dfrac{1}{2}$

Example 9

Work out the value of   a $8^{\frac{2}{3}}$ b $16^{-\frac{3}{4}}$

a $8^{\frac{2}{3}} = (8^{\frac{1}{3}})^2$
$= 2^2$ ← Use the rule $(a^m)^n = a^{mn}$. Work out the cube root of 8 first. Then square your answer.
$= 4$

b $16^{-\frac{3}{4}} = \dfrac{1}{16^{\frac{3}{4}}}$

$= \dfrac{1}{(16^{\frac{1}{4}})^3}$ ← Use $a^{-n} = \frac{1}{a^n}$.

$= \dfrac{1}{2^3}$

$= \dfrac{1}{8}$

ResultsPlus
Examiner's Tip

It is easier to work out the root first as this makes the numbers smaller and easier to manage.

B

Exercise 5F

1 Work out the value of the following.

a $9^{\frac{1}{2}}$ b $49^{\frac{1}{2}}$ c $100^{\frac{1}{2}}$ d $4^{\frac{1}{2}}$ e $\left(\frac{1}{4}\right)^{\frac{1}{2}}$

2 Work out the value of

a $27^{\frac{1}{3}}$ b $1000^{\frac{1}{3}}$ c $(-64)^{\frac{1}{3}}$ d $125^{\frac{1}{3}}$ e $\left(\frac{1}{8}\right)^{\frac{1}{3}}$

3 Work out the value of

a $16^{-\frac{1}{4}}$ b $4^{-\frac{1}{2}}$ c $125^{-\frac{1}{3}}$ d $\left(\frac{1}{32}\right)^{-\frac{1}{5}}$ e $\left(\frac{4}{9}\right)^{-\frac{1}{2}}$

A

4 Work out the value of

a $27^{\frac{2}{3}}$ b $1000^{\frac{2}{3}}$ c $64^{\frac{2}{3}}$ d $16^{\frac{3}{4}}$ e $25^{\frac{3}{2}}$

5 Work out, as a single fraction, the value of

a $125^{-\frac{2}{3}}$ b $10\,000^{-\frac{3}{4}}$ c $27^{-\frac{1}{3}}$ d $8^{-\frac{2}{3}}$ e $64^{-\frac{3}{2}}$

f $125^{-\frac{2}{3}} \times \left(\frac{1}{5}\right)^2$ g $8^{-\frac{1}{3}} \times \left(\frac{2}{5}\right)^2$

A★

6 Find the value of n

a $\frac{1}{8} = 8^n$ b $64 = 2^n$ c $\frac{1}{\sqrt{5}} = 5^n$ d $(\sqrt{7})^5 = 7^n$ e $(\sqrt[3]{2})^{11} = 2^n$

5.4 Using surds

◎ Objectives

● You can simplify surds.
● You can expand expressions involving surds.
● You can rationalise the denominator of a fraction.

❓ Why do this?

Surds occur in nature. The golden ratio $\frac{1+\sqrt{5}}{2}$ occurs in the arrangement of branches along the stems of plants, as well as veins and nerves in animal skeletons.

◈ Get Ready

1. Write down the first 10 square numbers.
2. Write down the value of   a $\sqrt{36}$ b $\sqrt{100}$
3. Which of these have an exact answer: $\sqrt{5}, \sqrt{9}, \sqrt{37}, \sqrt{64}$?

⬡ Key Points

◉ A number written exactly using square roots is called a **surd**.
$\sqrt{2}$ and $\sqrt{3}$ are both surds.

◉ $2 - \sqrt{3}$ and $5 + \sqrt{2}$ are examples of numbers written in surd form.
$\sqrt{4}$ is not a surd as $\sqrt{4} = 2$.

◉ These two rules can be used to simplify surds.
$$\sqrt{m} \times \sqrt{n} = \sqrt{mn} \qquad \frac{\sqrt{m}}{\sqrt{n}} = \sqrt{\frac{m}{n}}$$

◉ Simplified surds should never have a surd in the denominator.

● To **rationalise the denominator** of a fraction means to get rid of any surds in the denominator.

● To rationalise the denominator of $\dfrac{a}{\sqrt{b}}$ you multiply the fraction by $\dfrac{\sqrt{b}}{\sqrt{b}}$. This ensures that the final fraction has an integer as the denominator.

$$\frac{a}{\sqrt{b}} = \frac{a}{\sqrt{b}} \times \frac{\sqrt{b}}{\sqrt{b}} = \frac{a \times \sqrt{b}}{\sqrt{b} \times \sqrt{b}} = \frac{a\sqrt{b}}{b}$$

Example 10   Simplify $\sqrt{12}$.

$\sqrt{12} = \sqrt{4 \times 3}$

$\quad = \sqrt{4} \times \sqrt{3}$ ←   Use $\sqrt{m} \times \sqrt{n} = \sqrt{mn}$. $\sqrt{4} = 2$.

$\quad = 2\sqrt{3}$

Example 11   Expand and simplify $(2 + \sqrt{3})(4 + \sqrt{3})$.

$(2 + \sqrt{3})(4 + \sqrt{3}) = 8 + 2\sqrt{3} + 4\sqrt{3} + \sqrt{3} \times \sqrt{3}$ ←   Multiply out the brackets.

$\quad\quad = 8 + 6\sqrt{3} + 3$ ←   Simplify the expression.

$\quad\quad = 11 + 6\sqrt{3}$

Exercise 5G

1. Find the value of the integer k.
   a $\sqrt{8} = k\sqrt{2}$ b $\sqrt{18} = k\sqrt{2}$ c $\sqrt{50} = k\sqrt{2}$ d $\sqrt{80} = k\sqrt{5}$

2. Simplify
   a $\sqrt{200}$ b $\sqrt{32}$ c $\sqrt{20}$ d $\sqrt{28}$

3. Solve the equation $x^2 = 30$, leaving your answer in surd form.

4. Expand these expressions. Write your answers in the form $a + b\sqrt{c}$ where a, b and c are integers.
   a $\sqrt{3}(2 + \sqrt{3})$ b $(\sqrt{3} + 1)(2 + \sqrt{3})$ c $(\sqrt{5} - 1)(2 + \sqrt{5})$
   d $(\sqrt{7} + 1)(2 - \sqrt{7})$ e $(2 - \sqrt{3})^2$ f $(\sqrt{2} + 5)^2$

5. The area of a square is 40 cm². Find the length of one side of the square.
   Give your answer as a surd in its simplest form.

6. The lengths of the sides of a rectangle are $(3 + \sqrt{5})$ cm and $(3 - \sqrt{5})$ cm.
   Work out, in their simplified forms:
   a the perimeter of the rectangle   b the area of the rectangle.

7. The length of the side of a square is $(1 + \sqrt{2})$ cm. Work out the area of the square.
   Give your answer in the form $(a + b\sqrt{2})$ cm² where a and b are integers.

A

A*

Example 12 Rationalise the denominator of $\dfrac{2}{\sqrt{3}}$.

$$\frac{2}{\sqrt{3}} = \frac{2}{\sqrt{3}} \times \frac{\sqrt{3}}{\sqrt{3}} \quad \longleftarrow \quad \boxed{\text{Multiply the fraction by } \dfrac{\sqrt{3}}{\sqrt{3}}.}$$

$$= \frac{2 \times \sqrt{3}}{\sqrt{3} \times \sqrt{3}} \quad \longleftarrow \quad \boxed{\begin{array}{l}\text{Simplify the denominator by using}\\ \text{the fact that } \sqrt{3} \times \sqrt{3} = 3.\end{array}}$$

$$= \frac{2\sqrt{3}}{3}$$

Example 13 Rationalise the denominator of $\dfrac{15 - \sqrt{5}}{\sqrt{5}}$ and give your answer in the form $a + b\sqrt{5}$.

$$\frac{15 - \sqrt{5}}{\sqrt{5}} = \frac{15 - \sqrt{5}}{\sqrt{5}} \times \frac{\sqrt{5}}{\sqrt{5}}$$

$$= \frac{15\sqrt{5} - \sqrt{5} \times \sqrt{5}}{\sqrt{5} \times \sqrt{5}}$$

$$= \frac{15\sqrt{5} - 5}{5} \quad \longleftarrow \quad \boxed{\begin{array}{l}\text{Simplify the fraction by dividing both parts of}\\ \text{the expression on the top of the fraction by 5.}\end{array}}$$

$$= -1 + 3\sqrt{5}$$

ResultsPlus

Watch Out!

Remember to multiply both parts of the expression on the top of the fraction.

Exercise 5H

A

1 Rationalise the denominators and simplify your answers, if possible.

   a $\dfrac{1}{\sqrt{2}}$ **b** $\dfrac{1}{\sqrt{5}}$ **c** $\dfrac{5}{\sqrt{10}}$ **d** $\dfrac{2}{\sqrt{2}}$ **e** $\dfrac{4}{\sqrt{12}}$

A*

2 Rationalise the denominators and give your answers in the form $a + b\sqrt{c}$ where a, b and c are integers.

   a $\dfrac{2 + \sqrt{2}}{\sqrt{2}}$ **b** $\dfrac{6 - \sqrt{2}}{\sqrt{2}}$ **c** $\dfrac{10 + \sqrt{5}}{\sqrt{5}}$ **d** $\dfrac{12 - \sqrt{3}}{\sqrt{3}}$ **e** $\dfrac{14 + \sqrt{7}}{\sqrt{7}}$

3 The diagram shows a right-angled triangle.
The lengths are given in centimetres.
Work out the area of the triangle.
Give your answer in the form $a + b\sqrt{c}$ where a, b and c are integers.

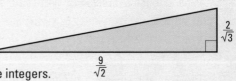

4 Solve these equations leaving your answers in surd form.

   a $x^2 - 6x + 2 = 0$ **b** $x^2 + 10x + 14 = 0$

5 The diagram represents a right-angled triangle ABC.
AB $= (\sqrt{7} + 2)$ cm   AC $= (\sqrt{7} - 2)$ cm.
Work out, leaving any appropriate answers in surd form:

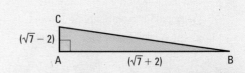

   a the area of triangle ABC

   b the length of BC, given that BC $= \sqrt{(AC^2 + AB^2)}$.

Chapter review

- For non-zero values of a
  $$a^0 = 1$$
- For any number n
  $$a^{-n} = \frac{1}{a^n}$$
- **Standard form** is used to represent very large or very small numbers.
- A number is in standard form when it is in the form $a \times 10^n$ where $1 \leqslant a < 10$ and n is an integer.
- It is often easier to multiply and divide very large or very small numbers, or estimate a calculation, if the numbers are written in standard form.
- Indices can be fractions. In general,
  $$a^{\frac{1}{n}} = \sqrt[n]{a}$$
- A number written exactly using square roots is called a **surd**.
- These two laws can be used to simplify surds.
  $$\sqrt{m} \times \sqrt{n} = \sqrt{mn} \qquad \frac{\sqrt{m}}{\sqrt{n}} = \sqrt{\frac{m}{n}}$$
- Simplified surds should never have a surd in the denominator.
- To **rationalise the denominator** of a fraction means to get rid of any surds in the denominator.
- To rationalise the denominator of $\frac{a}{\sqrt{b}}$ you multiply the fraction by $\frac{\sqrt{b}}{\sqrt{b}}$. This ensures that the final fraction has an integer as the denominator.

Review exercise

1 Write these as ordinary numbers.
- **a** 4×10^3
- **b** 2.6×10^8
- **c** 3.7×10^2
- **d** 9×10^{-4}
- **e** 1.35×10^{-1}
- **f** 8.001×10^{-5}

2 Work out the values of
- **a** 4^0
- **b** 4^{-1}
- **c** 2^0
- **d** 2^{-3}

3 Work out the values of
- **a** 3^0
- **b** $(-3)^0$
- **c** 3^{-1}
- **d** $\left(\frac{1}{3}\right)^0$

4 Work out the values of
- **a** $\frac{1}{3^{-1}}$
- **b** $\left(\frac{1}{3}\right)^{-1}$
- **c** 2×4^{-1}
- **d** $\frac{2}{4^{-1}}$

5 Work out
- **a** $9^{\frac{1}{2}}$
- **b** $100^{\frac{1}{2}}$
- **c** $8^{\frac{1}{3}}$
- **d** $64^{\frac{1}{3}}$

6 Work out
- **a** $9^{-0.5}$
- **b** $49^{-\frac{1}{2}}$
- **c** $125^{-\frac{1}{3}}$
- **d** $8^{-\frac{1}{3}}$

7 Work out
- **a** $4^{\frac{1}{2}}$
- **b** $8^{-\frac{1}{3}}$

June 2009

B

a i Write 7900 in standard form **ii** Write 0.000 35 in standard form.

b Work out $\dfrac{4 \times 10^3}{8 \times 10^{-5}}$ Give your answer in standard form.

9 In 2003 the population of Great Britain was 6.0×10^7.
In 2003 the population of India was 9.9×10^8.
Work out the difference between the population of India and the population of Great Britain in 2003.
Give your answer in standard form.

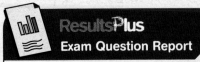

ResultsPlus
Exam Question Report

80% of students answered this question well.
They knew how to use their calculators properly.

June 2007

10 $3 \times \sqrt{27} = 3^n$ Find the value of n. *June 2006*

11 $8\sqrt{8}$ can be written in the form 8^k.

a Find the value of k.

$8\sqrt{8}$ can also be expressed in the form $m\sqrt{2}$ where m is a positive integer.

b Find the value of m.

c Rationalise the denominator of $\dfrac{1}{8\sqrt{8}}$.

Give your answer in the form $\dfrac{\sqrt{2}}{p}$ where p is a positive integer. *June 2006*

12 Solve

a $4^x = \dfrac{1}{16}$ **b** $2^x = \dfrac{1}{16}$ **c** $2 \times 2^{-x} = \dfrac{1}{4}$ **d** $2^{2x} = \dfrac{1}{2}$

13 Calculate $\dfrac{1}{\sqrt{2}+1} + \dfrac{1}{\sqrt{3}+\sqrt{2}} + \dfrac{1}{\sqrt{4}+\sqrt{3}} + \ldots \ldots \dfrac{1}{10+\sqrt{99}}$

6 RATIO

The ratio of water to land on the Earth's surface is approximately 7 : 3. However, this ratio is likely to change as global warming leads to the melting of the polar icecaps and an increase in the amount of the surface underwater. Countries including Bangladesh, Burma and Egypt could find large parts of their surface area flooded and even parts of the UK such as Lincolnshire and the area around the Thames Estuary may be lost.

⊙ Objectives

In this chapter you will:
- simplify ratios
- solve problems using ratios
- share a quantity in a given ratio.

◑ Before you start

You should be able to:
- find the HCF of two numbers
- use ratios in maps and scale drawings
- convert between metric units.

6.1 Introducing ratio

⊙ Objectives

○ You can simplify ratios.

○ You can write down a fraction from a ratio.

○ You can write ratios in the unitary form.

⑦ Why do this?

Maps use ratio so that the actual distance between two places can be worked out by measuring the distance on the map and then using the ratio.

⬆ Get Ready

1. The scale on a road map is 1 : 200 000.
   Sunderland and Newcastle are 9 cm apart on the map.
   Work out the real distance, in km, between Sunderland and Newcastle.

🔍 Key Points

◉ **Ratios** are used to compare quantities.

◉ The simplest form of a ratio has whole numbers with no common factor.

◉ Ratios are sometimes given in the form $1 : n$ where n is a number.
   This is called the **unitary** form of a ratio.
   It is most often used for scales in maps and scale drawings (see Unit 3 Chapter 14).

◉ To write a ratio in the form $1 : n$, divide each number in the ratio by the first number in the ratio.
   For example, $5 : 8 = \frac{5}{5} : \frac{8}{5}$
   $$= 1 : 1.6$$

🔍 Example 1

In a library, there are **560** fiction books and **420** non-fiction books.

a Write down the ratio of the number of fiction books to the number of non-fiction books.
Give your ratio in its simplest form.

b Give your ratio in the form $1 : n$.

a $560 : 420$ ⟵ Write down the ratio. The number of fiction books goes first.
$= 56 : 42$ Divide both numbers by 10.
$= 4 : 3$ Divide both numbers by 14.
4 : 3 is the simplest form.

b $4 : 3 = 1 : \frac{3}{4}$ ⟵ Divide both numbers by 4 to give the ratio in the form $1 : n$.
(or $1 : 0.75$)

⚙ Exercise 6A

Questions in this chapter are targeted at the grades indicated.

1 Write each ratio in its simplest ratio.
   a $2 : 8$ b $10 : 6$ c $49 : 63$ d $121 : 44$

D

2 Write each ratio in the form $1 : n$.
   a $5 : 15$ b $8 : 32$ c $4 : 14$ d $6 : 3$
   e $30 : 9$ f $15 : 9$ g $\frac{1}{2} : 4$ h $\frac{1}{4} : \frac{2}{10}$

3 In a school, there are 120 computers. There are 600 students in the school.
   Advise the Headteacher of the ratio of the number of computers to the number of stu[...]
   Give your ratio in the form $1 : n$.

4 In a cinema, there are 160 children and 200 adults.
   a What fraction of the audience are children?
   b Write down the ratio of the number of children to the number of adults.
      Give your ratio in its simplest form.
   c Write your answer to part b in the form $1 : n$.

5 The length of a model aeroplane is 16 cm. The length of the real aeroplane is 60 m.
   Work out the ratio of the length of the model aeroplane to the length of the real aeroplane.
   Write your answer in the form $1 : n$.

6 Write these ratios in the form $1 : n$.
   a 3 hours : $\frac{1}{2}$ hour   b £2 : 40p   c 2 m : 4 cm   d 25 g : 1 kg

6.2 Solving ratio problems

Objective

- You can solve problems using ratio.

Why do this?

A teacher taking some pupils on a school trip knows that the ratio of staff to students must be 1 : 15. Once the number of pupils on the trip is known, the number of staff needed can be calculated.

Get Ready

1. Write the ratio in its simplest form.
    a 10 : 15   b 130 : 650   c 4 cm : 35 mm   d 45 g : 1 kg

Key Point

- If the ratio of two quantities is given and one of the quantities is known, then the other quantity can be found. This can be done using **equivalent ratios**.

Example 2

To make concrete, 2 parts of cement is used to every 5 parts of sand.

   a Write down the ratio of cement to sand.
   b 4 buckets of cement are used. How many buckets of sand will be needed?
   c 20 buckets of sand are used. How many buckets of cement will be needed?

a 2 : 5

b $4 \div 2 = 2$ ← *The amount of cement has been multiplied by 2.*

cement : sand

$\times 2 \left(\begin{array}{ccc} 2 & : & 5 \\ 4 & : & 10 \end{array} \right) \times 2$ ← *Multiply 5 by 2.*

10 buckets of sand will be needed.

$20 \div 5 = 4$

← The amount of sand has been multiplied by 4.

cement : sand

$\times 4 \Big(\begin{array}{ccc} 2 & : & 5 \\ 8 & : & 20 \end{array} \Big) \times 4$

← Multiply 2 by 4.

8 buckets of cement will be needed.

Exercise 6B

1. In a recipe for pancakes, the ratio of the weight of flour to the weight of sugar is 4 : 1
   Work out the weight of sugar needed for:
   a 40 g of flour
   b 120 g of flour
   c 1 kg of flour.

2. Brass is made from copper and zinc in the ratio 5 : 3 by weight.
   a If there are 6 kg of zinc, work out the weight of copper.
   b If there are 25 kg of copper, work out the weight of zinc.

3. A map is drawn using a scale of 1 : 500 000. On the map, the distance between two towns is 21.7 cm.
   Work out the real distance between the towns. Give your answer in kilometres.

4. George and Henry share some money in the ratio 7 : 9
   If George receives £840, work out how much money Henry gets.

5. The ratio of the widths of two pictures is 6 : 9
   If the width of the first picture is 1.02 m, calculate the width of the second picture.

6. In a school, the ratio of the number of students to the number of computers is $1 : \frac{2}{5}$.
   If there are 100 computers in the school, work out the number of students in the school.

6.3 Sharing a quantity in a given ratio

Objective

- You can share a quantity in a given ratio.

Why do this?

If a recipe required 300 g of crumble mixture and you know that the ratio of sugar, fat and flour is 1 : 2 : 3, you could work out how much of each you would need.

Get Ready

Work out 1. $\frac{16.52}{4} \times 5$ 2. $8.4 \times \frac{5}{7}$ 3. $\frac{483}{6} \times 3$

Key Point

- There are two methods for sharing a quantity in a given ratio. In one method, you work out how much each share is worth, then multiply by the number of shares each person receives. In the second method, you work out what fraction of the total amount each person receives and multiply the total by these fractions.

Example 3

Anna, Faye and Harriet share £42 in the ratio 1 : 2 : 3
How much money does each girl get?

Method 1

$1 + 2 + 3 = 6$ ← Add the numbers in the ratio to get the total number of shares.

$42 \div 6 = 7$ ← Work out what each share is worth.

Anna gets £7. ← Anna gets 1 share.

Faye gets $7 \times 2 = £14$. ← Faye gets 2 shares so multiply 7 by 2.

Harriet gets $7 \times 3 = £21$. ← Harriet gets 3 shares so multiply 7 by 3.

ResultsPlus
Examiner's Tip

Check your answer is correct by adding up each person's share and check this equals the total number of shares.

Method 2

Anna receives $\frac{1}{6}$ of the total amount

Faye receives $\frac{2}{6} = \frac{1}{3}$

Harriet receives $\frac{3}{6} = \frac{1}{2}$

Add each number in the ratio to find the denominator.

Anna $\frac{1}{6} \times £42 = \frac{\overset{7}{\cancel{42}}}{\cancel{6}} = £7$

Faye $\frac{1}{3} \times £42 = \frac{\overset{14}{\cancel{42}}}{\cancel{3}} = £14$

Harriet $\frac{1}{2} \times £42 = \frac{\overset{21}{\cancel{42}}}{\cancel{2}} = £21$

Exercise 6C

1 Divide the quantities in the ratios given.
   a £14.91 in the ratio 2 : 5
   b 600 g in the ratio 3 : 2
   c £170.52 in the ratio 1 : 4 : 7
   d 34.65 m in the ratio 2 : 4 : 5

2 The angles in a triangle are in the ratio 6 : 5 : 7
   Find the sizes of the three angles.

3 Three boys washed some cars. They earned a total of £87.60.
   They shared the money in the ratio of the amount of time that each of them worked.
   James worked for 5 hours. Sam worked for $3\frac{1}{2}$ hours and Will also worked for $3\frac{1}{2}$ hours.
   Calculate the amount of money James received.

   A02
   A03

4 Jean and Kevin shared £320 in the ratio 3 : 5
   Jean gave one third of her share to Michael.
   Kevin gave half of his share to Michael.
   What fraction of the original amount of money did Michael receive?
   Give your fraction in its simplest form.

   A03

C

C A03

5 Barry bought a box full of fruit. The box contained some apples, oranges and lemons in the ratio 5 : 3 : 1
Given that there were more than 50 pieces of fruit in the box, work out the minimum number of oranges in the box.

A03

6 Angela and Michelle shared some money in the ratio 4 : 9
Then Angela gave Daniel half of her share.
Michelle gave Daniel a third of her share.
Daniel was given a total of £20.
Work out how much money was shared originally by Angela and Michelle.

Chapter review

- **Ratios** are used to compare quantities.
- The simplest form of a ratio has whole numbers with no common factor.
- Ratios are sometimes given in the form $1 : n$ where n is a number. This is called the **unitary** form of a ratio.
- To write a ratio in the form $1 : n$, divide each number in the ratio by the first number in the ratio.
- If the ratio of two quantities is given and one of the quantities is known, then the other quantity can be found. This can be done using **equivalent ratios**.
- There are two methods for sharing a quantity in a given ratio. Either work out how much each share is worth and multiply by the number of shares each person receives, or work out what fraction of the total amount each person receives and multiply the total by these fractions.

Review exercise

1 There are some sweets in a bag.
18 of the sweets are toffees.
12 of the sweets are mints.
Write down the ratio of the number of toffees to the number of mints.
Give your ratio in its simplest form.

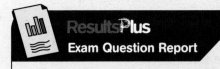

Results**Plus**
Exam Question Report

79% of students answered this question well because they displayed their answers in the form asked for in the question.

June 2009

2 A coin is made from copper and nickel.
84% of its weight is copper.
16% of its weight is nickel.
Find the ratio of the weight of copper to the weight of nickel.
Give your answer in its simplest form.

June 2008

D

3 Write each ratio in the form $1 : n$.
a 18 : 6   b 3 : 42   c 68 : 4   d $\frac{1}{6} : \frac{2}{5}$

D

4 The distance from Ailing to Beeford is 2 km. The distance from Ceetown to Deeton is 800 metres.
Write the following as a ratio.
Distance from Ailing to Beeford : Distance from Ceetown to Deeton
Give your answer in its simplest form.

5 Alice builds a model of a house. She uses a scale of 1 : 20
The height of the real house is 10 metres.
a Work out the height of the model.
The width of the model is 80 cm.
b Work out the width of the real house.

C

6 Mr Brown makes some compost.
He mixes soil, manure and leaf mould in the ratio 3 : 1 : 1
Mr Brown makes 75 litres of compost.
How many litres of soil does he use?

Nov 2006

7 A garage sells British cars and foreign cars.
The ratio of the number of British cars sold to the number of foreign cars sold is 2 : 7.
The garage sells 45 cars in one week.
Work out the number of British cars the garage sold that week.

June 2008

8 There are 600 counters in a bag.
90 of the 600 are yellow. 180 of the 600 are red.
The rest of the counters in the bag are blue or green.
There are twice as many blue counters as green counters.
Work out the number of green counters in the bag.

A03

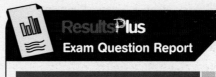

Results Plus
Exam Question Report

72% of students answered this sort of question poorly because they did not work out the total number of shares first.

May 2009

A02
A03

*** 9** Robert wants to buy some new golf clubs.
He is considering buying them from the USA over the internet instead of from his local golf professional.
Use the prices quoted below to find which option is cheaper. Use the exchange rate £1 = $1.50.
Show all of your working.

| *Local professional* | *Imported from USA* |
| *£435* | *$570 plus taxes and duties of 20%* |

*** 10** Which bottle of tomato ketchup gives better value for money?
Show all your calculations.

A02
A03

720 g
£1.79

460 g
£1.00

11 A map is drawn to a scale of 1 : 50 000. A field is in the shape of a rectangle on the map. The area on the map is 6 cm². Work out the area of the field in real-life. Give your answer in km².

B

7 EXPRESSIONS AND SEQUENCES

Neptune was the first planet to be found by mathematical prediction. Scientists looked at the number patterns of the orbits of planets in the Solar System and correctly predicted Neptune's position to within a degree. Using the predicted position, Johann Galle identified Neptune almost immediately on 23 September 1846.

◎ Objectives

In this chapter you will:
- distinguish the different roles played by letter symbols in algebra and use the correct notation to derive algebraic expressions
- collect like terms
- use substitution to work out the value of an expression
- use the index laws applied to simple algebraic expressions and to algebraic expressions with fractional or negative powers
- generate terms of a sequence using term-to-term and position-to-term definitions
- derive and use the nth term of a sequence.

◇ Before you start

You need to be able to:
- simplify an expression where each term is in the same unknown or unknowns
- use directed numbers in calculations
- use index laws with numbers.

7.1 Collecting like terms

◉ Objectives

- ○ You can distinguish the different roles played by letter symbols in algebra and use the correct notation.
- ○ You can manipulate algebraic expressions by collecting like terms.

◈ Why do this?

Waitresses use algebra to note people's orders and then collect like terms to make the order simple for the chef.

⬆ Get Ready

Simplify   **1.** $a + a + a + a$ **2.** $4c - c + 5c$ **3.** $3p^2 - 5p^2 + 4p^2$

◔ Key Points

- ◉ $2x$, $3y$ and $2x + 3y$ are called **algebraic expressions**
- ◉ Each part of an **expression** is called a **term** of the expression. $2x$ and $3y$ are terms of the expression $2x + 3y$.
- ◉ When adding or subtracting expressions, different letter symbols cannot be combined. For example $2x + 3y$ cannot be simplified further.
- ◉ The sign of a term in an expression is always written before the term.
  For example, in the expression $4 + 2x - 3y$ the '+' sign means add $2x$ and the '−' sign means subtract $3y$.
- ◉ The term x can be written as $1x$.
- ◉ In algebra, BIDMAS describes the order of operations when **collecting like terms** (see Section 1.3 for use of BIDMAS).

Example 1   Simplify the expression $4p - 2q + 1 - 3p + 5q$.

$$4p - 2q + 1 - 3p + 5q$$
$$= 4p - 3p - 2q + 5q + 1$$
$$= p + 3q + 1$$

$-2 + 5 = +3$
so $-2q + 5q = +3q$

$4p - 3p = 1p$ which is written as just p.

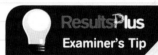

Results Plus
Examiner's Tip

Rewrite each expression with the like terms next to each other.

Example 2   Alfie is n years old. Bilal is 3 years older than Alfie. Carla is twice as old as Alfie.
Write down an expression, in terms of n, for the total of their ages in years.
Give your answer in its simplest form.

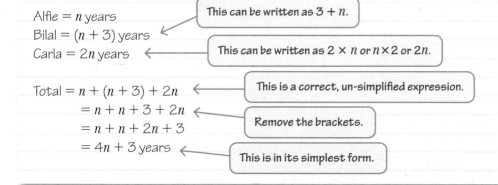

Alfie $= n$ years
Bilal $= (n + 3)$ years
Carla $= 2n$ years

This can be written as $3 + n$.

This can be written as $2 \times n$ or $n \times 2$ or $2n$.

Total $= n + (n + 3) + 2n$
       $= n + n + 3 + 2n$
       $= n + n + 2n + 3$
       $= 4n + 3$ years

This is a correct, un-simplified expression.

Remove the brackets.

This is in its simplest form.

Exercise 7A

Questions in this chapter are targeted at the grades indicated.

1 Simplify

a $5x + 2x + 3y + y$ b $3w + 7w + 4z - 2z$ c $3p + q + p + 4q$

d $4a + 3b - a - 2b$ e $c + 2d + 5c - 4d$ f $3m - 7n - m + 4n$

g $5e - 3f - e - 4f$ h $2x + 8y - 3 + 2y + 5$ i $3p - q + 2 - 5p + 4q - 7$

j $9 + a - 2b - 5a + 4 - 3b$

2 Georgina, Samantha and Mason collect football stickers. Georgina has x stickers in her collection. Samantha has 9 stickers less than Georgina. Mason has 3 times as many stickers as Georgina. Write down an expression, in terms of x, for the total number of these stickers. Give your answer in its simplest form.

A03 3 The diagram shows a triangle. Write down an expression, in terms of x and y, for the perimeter of this triangle. Give your answer in its simplest form.

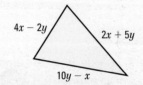

7.2 **Using substitution**

◉ **Objective**

○ Given the value of each letter in an expression, you can work out the value of the expression by substitution.

◈ **Why do this?**

In your science lessons you need to be able to substitute into formulae when carrying out many calculations.

⬆ **Get Ready**

Write expressions, in terms of x and y, for the perimeter of these rectangles:

1. length $2x + 4$, width $y + 2$ **2.** length $3y + 3$, width $x - 5$ **3.** length $4x + 5$, width $y - 2$.

🜨 **Key Point**

◉ If you are given the value for each letter in an expression then you can **substitute** the values into the expression and **evaluate** the expression.

🔑 **Example 3**   Work out the value of each of these expressions when $a = 5$ and $b = -3$.

a $4a + 3b$ b $a - 2b - 8$ c $2a^2 + 4b$

ResultsPlus
Examiner's Tip

Replace each letter with its numerical value.

a $4a + 3b = 4 \times 5 + 3 \times (-3)$
  $= 20 - 9$ ← Positive × negative = negative.
  $= 11$

b $a - 2b - 8 = 5 - 2 \times (-3) - 8$
    $= 5 + 6 - 8$
    $= 3$

Work out the multiplication first (BIDMAS). Negative × negative = positive.

c $2a^2 + 4b = 2 \times (5)^2 + 4 \times (-3)$
    $= 2 \times 25 - 12$
    $= 50 - 12$ ← It is only the value of a (=5) that is squared.
    $= 38$

Exercise 7B

1 Work out the value of each of these expressions when $x = 4$ and $y = -1$.

a $x + 3y$ b $x - y$ c $2x - 5y + 3$ d $4x + 1 + 2y$

2 Work out the value of each of these expressions when $p = -2$, $q = 3$ and $r = -5$.

a $p + q + r$ b $2q + 3r + 5p$ c $2q - r + 3p$

d $6 - q - 2r + p$ e $5p + 3q^2$ f $p^2 - 2q^2 + r^2$

7.3 Using the index laws

Objective

○ You understand and can use the index laws applied to simple algebraic expressions.

Why do this?

To write large numbers, like the speed of sound, indices are often used to shorten the way the value is written.

Get Ready

1. Write as a power of a single number.

a $4^3 \times 4^8$ b $\dfrac{7^8 \times 7^4}{7^5}$ c $(6^3)^2$

Key Point

◉ You can use the laws of indices to simplify algebraic expressions. See Section 1.4 for the index laws.

Example 4

a Simplify $c^3 \times c^4$

b Simplify $5y^3z^5 \times 2y^2z$

ResultsPlus
Watch Out!

Group like terms together before attempting to use the laws of indices.

a $c^3 \times c^4 = c \times c \times c \times c \times c \times c \times c$

$= c^7$ ← Note: $3 + 4 = 7$.

b $5y^3z^5 \times 2y^2z = 5 \times y^3 \times z^5 \times 2 \times y^2 \times z$ ← z is the same as z^1

$= 5 \times 2 \times y^3 \times y^2 \times z^5 \times z^1$

$= 10 \times y^{3+2} \times z^{5+1}$

$= 10 \times y^5 \times z^6$ Using $x^p \times x^q = x^{p+q}$

$= 10y^5z^6$

Exercise 7C

1 Simplify

a $m \times m \times m \times m \times m$ b $2p \times 3p$ c $q \times 4q \times 5q$

2 Simplify

a $a^4 \times a^7$ b $n \times n^3$ c $x^5 \times x$ d $y^2 \times y^3 \times y^4$

3 Simplify

a $2p^2 \times 6p^4$ b $4a \times 3a^4$ c $b^7 \times 5b^2$ d $3n^2 \times 6n$

4 Simplify

a $5t^3u^2 \times 4t^5u^3$ b $2xy^3 \times 3x^5y^4$ c $a^2b^5 \times 7a^3b$

d $4cd^5 \times 2cd^4$ e $2mn^2 \times 3m^3n^2 \times 4m^2n$

Example 5   a Simplify $d^5 \div d^2$

b Simplify $\dfrac{10x^2y^5}{2xy^3}$

a $d^5 \div d^2 = \dfrac{d^5}{d^2} = \dfrac{d \times d \times d \times d \times d}{d \times d}$

$= d^3$ ← Note: 5 − 2 = 3

ResultsPlus
Examiner's Tip

Write fractions, such as $\dfrac{p^5}{p^3}$ as $p^5 \div p^3$.

b $\dfrac{10x^2y^5}{2xy^3}$ is the same as $10x^2y^5 \div 2xy^3$

$10x^2y^5 \div 2xy^3 = (10 \div 2) \times (x^2 \div x) \times (y^5 \div y^3)$

$= 5 \times x^{2-1} \times y^{5-3}$

$= 5 \times x \times y^2$ Using $x^p \div x^q = x^{p-q}$

$= 5xy^2$

Exercise 7D

1 Simplify

a $a^7 \div a^4$ b $b^5 \div b$ c $\dfrac{c^8}{c^5}$ d $d^4 \div d^3$

2 Simplify

a $6q^5 \div 3q^3$ b $12p^7 \div 4p^2$ c $8x^6 \div 2x^5$ d $\dfrac{20y^8}{2y}$

3 Simplify

a $15a^5b^6 \div 3a^3b^2$ b $30p^3q^4 \div 6p^2q$ c $\dfrac{8c^4d^7}{2c^2d^3}$ d $\dfrac{6x^3 \times 2x^4}{4x^2}$

e $\dfrac{5m^2n \times 4mn^2}{2mn^2}$

Example 6 Simplify $(2c^3d)^4$

Method 1

$(2c^3d)^4 = (2)^4 \times (c^3)^4 \times (d)^4$

$\qquad = 16 \times c^{3 \times 4} \times d^{1 \times 4}$

$\qquad = 16 \times c^{12} \times d^4$

$\qquad = 16c^{12}d^4$

Using $(x^p)^q = x^{p \times q}$

ResultsPlus
Examiner's Tip

You must apply the power to number terms as well as the algebraic terms.

Method 2

$(2c^3d)^4$ can be written as $2c^3d \times 2c^3d \times 2c^3d \times 2c^3dd$

$= 2 \times 2 \times 2 \times 2 \times c^3 \times c^3 \times c^3 \times c^3 \times d \times d \times d \times d$

$= 16 \times c^{3+3+3+3} \times d^4$

$= 16 \times c^{12} \times d^4$

Using $x^p \times x^q = x^{p+q}$

$= 16c^{12}d^4$

Exercise 7E

1. Simplify
   a $(a^7)^2$
   b $(b^3)^5$
   c $(c^3)^3$
   d $(d^2)^8$

2. Simplify
   a $(2p^3)^2$
   b $(3q^2)^4$
   c $(5x^4)^2$
   d $\left(\dfrac{m^4}{2}\right)^3$

3. Simplify
   a $(2x^3y^2)^4$
   b $(7e^5f^3)^2$
   c $(5p^5q)^3$
   d $\left(\dfrac{2x^4y^2}{3xy^4}\right)^3$

C

B

7.4 Fractional and negative powers

◉ Objective

● You can use the index laws applied to algebraic expressions with fractional or negative powers.

❓ Why do this?

To write very small numbers, like the radius of a molecule, negative powers of 10 are used.

⬆ Get Ready

Simplify these expressions.

1. $(a^3)^6$
2. $(3y^5)^3$
3. $\left(\dfrac{4a^3b^2}{2a^2b^5}\right)^2$

Key Points

⦿ The laws of indices used so far can be used to develop two further laws.

$x^4 \div x^4 = x^{4-4} = x^0$

Also

$x^4 \div x^4 = 1$ since any term divided by itself is equal to 1.

Therefore $x^0 = 1$

In general

$x^0 = 1$

$x^3 \div x^4$

$= \dfrac{x \times x \times x}{x \times x \times x \times x} = \dfrac{1}{x}$

Also, using $x^p \div x^q = x^{p-q}$

$x^3 \div x^4 = x^{3-4} = x^{-1}$

Therefore $x^{-1} = \dfrac{1}{x}$

In general

$x^{-m} = \dfrac{1}{x^m}$

⦿ The laws of indices can be used further to solve problems with fractional indices.

The square root of x is written $\sqrt{x}$, and you know that:

$\sqrt{x} \times \sqrt{x} = x$

Using $x^p \times x^q = x^{p+q}$

$x^{\frac{1}{2}} \times x^{\frac{1}{2}} = x^{\frac{1}{2} + \frac{1}{2}} = x^1 = x$

and so, $x^{\frac{1}{2}} = \sqrt{x}$

Also, $x^{\frac{1}{3}} \times x^{\frac{1}{3}} \times x^{\frac{1}{3}} = x$, showing that $x^{\frac{1}{3}} = \sqrt[3]{x}$

In general

$x^{\frac{1}{n}} = \sqrt[n]{x}$

Example 7 Simplify $(3x^4y)^{-2}$

$(3x^4y)^{-2} = \dfrac{1}{(3x^4y)^2}$ ← Using $x^{-m} = \dfrac{1}{x^m}$

$= \dfrac{1}{9x^8y^2}$ ← Using $(x^p)^q = x^{p \times q}$

ResultsPlus

Examiner's Tip

Remember that a negative power just means 'one over' or 'the reciprocal of'.

Exercise 7F

B

1 Simplify

  a a^{-1} b $(b^2)^{-1}$ c c^{-2} d $(d^3)^{-1}$

2 Simplify

  a $(e^3)^{-2}$ b $(f^2)^{-4}$ c $(x^{-1})^{-2}$ d $(y^{-1})^{-1}$

A

3 Simplify

  a $(x^2y^7)^0$ b $(2x^4y^5)^0$ c $(5p^2q^4)^{-1}$ d $(3c^3d)^{-3}$

  e $\left(\dfrac{2p^3q}{3r^2}\right)^{-2}$

Simplify $(8x^6y^4)^{\frac{1}{3}}$

$$(8x^6y^4)^{\frac{1}{3}} = 8^{\frac{1}{3}} \times (x^6)^{\frac{1}{3}} \times (y^4)^{\frac{1}{3}}$$

Using $x^{\frac{1}{n}} = \sqrt[n]{x}$

$$= \sqrt[3]{8} \times x^{6 \times \frac{1}{3}} \times y^{4 \times \frac{1}{3}}$$

Using $(x^p)^q = x^{p \times q}$

$$= 2 \times x^2 \times y^{\frac{4}{3}}$$

$$= 2x^2y^{\frac{4}{3}}$$

ResultsPlus
Examiner's Tip

Remember that the denominator of the index is the root.

Exercise 7G

1 Simplify
  a $(9a^4)^{\frac{1}{2}}$
  b $(16c^2)^{\frac{1}{4}}$
  c $(27e^3f^{-9})^{\frac{1}{3}}$
  d $(100x^3y^5)^{\frac{1}{2}}$

2 Simplify
  a $(a^4)^{-\frac{1}{2}}$
  b $(8c^3)^{-\frac{1}{3}}$
  c $(32x^9y^5)^{-\frac{1}{5}}$
  d $(x^2y^6)^{-\frac{1}{4}}$

A

7.5 Term-to-term and position-to-term definitions

◎ Objective

● You can generate terms of a sequence using term-to-term and position-to-term definitions of the sequence.

⑦ Why do this?

To recognise world trends in specific illnesses, patterns linking data are often used.

⬆ Get Ready

Continue these number patterns.
1. 2, 4, 6, 8, 10, …   **2.** 4, 9, 14, 19, 24, 29, …   **3.** 1, 3, 5, 7, 9, …

🔍 Key Points

● A **sequence** is a pattern of shapes or numbers which are connected by a **rule** (or definition of the sequence).
● The relationship between consecutive terms describes the rule which enables you to find subsequent **terms of the sequence**.

Here is a sequence of 4 square patterns made up of squares:

   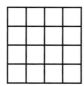

Pattern 1   Pattern 2   Pattern 3   Pattern 4

- Each pattern above is a term of the sequence;

  ☐ is the 1st term in the sequence,

  ⊞ is the 2nd term in the sequence, etc.

- The number of squares in each term form a sequence of numbers, 1, 4, 9, 16, …
- The odd numbers form a sequence, 1, 3, 5, …
- The even numbers form a sequence, 2, 4, 6, …
- You can continue a sequence if you know how the terms are related: the **term-to-term rule**.
- You can continue a sequence if you know how the position of a term is related to the definition of the sequence: the **position-to-term rule**.

Example 9 Find **a** the next term, and

b the 12th term of the sequence of numbers: 1, 4, 9, 16, …

1st term	2nd term	3rd term	4th term	5th term
1	4	9	16	

$+3 \qquad +5 \qquad +7$

> The difference between consecutive terms increases by 2.
> This is the term-to-term rule which enables you to find subsequent terms of the sequence.

a The difference between the 4th and the 5th term is $+9$ and so the 5th term is $16 + 9 = $ **25**.

b The 6th term $= 6^2 = 36$, the 7th term $= 7^2 = 49$, etc.

> The numbers 1 ($= 1^2$), 4 ($= 2^2$), 9 ($= 3^2$), 16 ($= 4^2$) and 25 ($= 5^2$) are the first five square numbers.

The 12th term $= 12^2 = $ **144**.

> In this way a term of the sequence can be found by the position of the term in the sequence.

Exercise 7H

Find **a** the term-to-term rule,

  b the next two terms, and

  c the 10th term for each of the following number sequences.

1	2	5	8	11
2	−4	2	8	14
3	19	12	5	−2
4	1	3	6	10
5	0	2	6	12

7.6 The nth term of an arithmetic sequence

⊙ Objectives

- ⊙ You can use linear expressions to describe the nth term of a sequence.
- ⊙ You can use the nth term of a sequence to generate terms of the sequence.

⊘ Why do this?

To be able to predict how many people might catch the flu, epidemiologists need to develop a general rule.

⬆ Get Ready

Find **a** the rule, **b** the next two terms, **c** the 10th term for each of the following number sequences.

1. 1, 4, 7, 10, …   **2.** −4, −1, 2, 5, 8, …   **3.** 124, 118, 112, 106, 100, …

🌰 Key Points

- ⊙ An **arithmetic sequence** is a sequence of numbers where the rule is simply to add a fixed number.
  For example, 2, 5, 8, 11, 14, … is an arithmetic sequence with the rule 'add 3'.
  In this example the fixed number is 3.
- ⊙ This is sometimes called the **difference** between consecutive terms.
- ⊙ You can find the nth term using the result nth term = n × difference + **zero term**.
- ⊙ You can use the nth term to generate the terms of a sequence.
- ⊙ You can use the terms of a sequence to find out whether or not a given number is part of a sequence, and explain why.

🔍 Example 10

Here are the first five terms of an arithmetic sequence: 2, 5, 8, 11, 14, …

a Write down, in terms of n, an expression for the nth term of the arithmetic sequence.

b Use your answer to part **a** to find the 20th term.

zero term	1st term	2nd term	3rd term	4th term	5th term
−1	2	5	8	11	14

$+3 \quad +3 \quad +3 \quad +3 \quad +3$

difference

a The zero term is the term before the first term.
Work out the zero term by using the difference of $+3$.
Zero term = $2 - 3 = -1$ ⟵ Inverse of $+3$.

The nth term = n × difference + zero term
nth term = $n \times +3 + -1$
$= 3n - 1$

b For the 20th term, $n = 20$
When $n = 20$, $3n - 1 = 3 \times 20 - 1$
$= 60 - 1 = 59$
So the 20th term is 59.

ResultsPlus
Examiner's Tip

Always check your answer by substituting values of n into your nth term.
For example,
1st term, when $n = 1$, $3n - 1 = 3 \times 1 - 1 = 2$ ✓
2nd term, when $n = 2$, $3n - 1 = 3 \times 2 - 1 = 5$ ✓
3rd term, when $n = 3$, $3n - 1 = 3 \times 3 - 1 = 8$ ✓
etc.

Exercise 7I

C

1 Write down **i** the difference between consecutive terms
   ii the zero term for each of the following arithmetic sequences.
   a $0, 2, 4, 6, 8, \ldots$ **b** $-7, -3, 1, 5, 9, \ldots$ **c** $14, 9, 4, -1, -6, \ldots$

2 Here are the first five terms of an arithmetic sequence: $1, 7, 13, 20, 26, \ldots$
   a Write down, in terms of n, an expression for the nth term of this arithmetic sequence.
   b Use your answer to part **a** to work out the **i** 12th term, **ii** 50th term.

3 Here are the first four terms of an arithmetic sequence: $7, 11, 15, 19, \ldots$
   a Write down, in terms of n, an expression for the nth term of this arithmetic sequence.
   b Use your answer to part **a** to work out the **i** 15th term, **ii** 100th term.

4 Here are the first five terms of an arithmetic sequence: $32, 27, 22, 17, 12, \ldots$
   a Write down, in terms of n, an expression for the nth term of this arithmetic sequence.
   b Use your answer to part **a** to work out the **i** 20th term, **ii** 200th term.

A03

5 Here are the first four terms of an arithmetic sequence: $18, 25, 32, 39, \ldots$
   Explain why the number 103 cannot be a term of this sequence.

A03 * **6** Here are the first five terms of an arithmetic sequence:
   7   11   15   19   23
   Pat says that 453 is a term in this sequence. Pat is wrong.
   Explain why.   Nov 2005

Chapter review

- $2x$, $3y$ and $2x + 3y$ are called **algebraic expressions**.
- Each part of an **expression** is called a **term** of the expression.
- When adding or subtracting expressions, different letter symbols cannot be combined.
- The sign of a term in an expression is always written before the term.
- The term x can be written as $1x$.
- In algebra, BIDMAS describes the order of operations when **collecting like terms**.
- If you are given the value for each letter in an expression then you can **substitute** the values into the expression and **evaluate** the expression.
- You can use the laws of indices to simplify algebraic expressions.
- The basic index laws can be used to develop further laws:
  $x^0 = 1$, for all values of x, $x^{-m} = \dfrac{1}{x^m}$ and $x^{\frac{1}{n}} = \sqrt[n]{x}$ where m and n are integers.
- A **sequence** is a pattern of shapes or numbers which are connected by a **rule** (or definition of the sequence).
- The relationship between consecutive terms describes the rule which enables you to find subsequent **terms of the sequence**.
- You can continue a sequence if you know how the terms are related: the **term-to-term rule**.
- You can continue a sequence if you know how the position of a term is related to the definition of the sequence: the **position-to-term rule**.
- An **arithmetic sequence** is a sequence of numbers where the rule is simply to add a fixed number. This is called the **difference** between consecutive terms.

⊙ You can find the nth term of an arithmetic sequence using the result nth term $= n \times$ difference $+$ **zero term**.

⊙ You can use the nth term of an arithmetic sequence to generate the terms of a sequence.

⊙ You can use the terms of a sequence to find out whether or not a given number is part of a sequence, and explain why.

Review exercise

1 Simplify   **a** $3x - 4y + 2x - y$ **b** $m - 7n + 5m - 3n$

2 Helen and Stuart collect stamps.

Helen has 240 British stamps and 114 Australian stamps.

  a Write down an algebraic expression that could be used to represent Helen's British and Australian stamps. Define the letters used.

Stuart has 135 British stamps and 98 Australian stamps.

  b Using the same letters, write down an algebraic expression that could be used to represent the total of Helen's and Stuart's British and Australian stamps.

3 Work out the value of each of these expressions when $x = 2$, $y = -3$ and $z = -7$

  a $3x + y$ **b** $x - 2y$ **c** $x + 3y - 2z$ **d** $5xy$ **e** $x^2 + y^2 + z^2$

4 Simplify

  a $y \times y \times y$ **b** $x \times 3x$ **c** $z^3 \times z^5$ **d** $p \times p^6$ **e** $2a^2 \times 8a^5$

5 Simplify

  a $a^6 \div a^3$ **b** $b^9 \div b^4$ **c** $21p^4 \div 3p$ **d** $\dfrac{24x^5}{3x^2}$ **e** $16a^6b^3 \div 2a^5b^3$

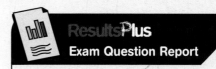

ResultsPlus

Exam Question Report

93% of students answered this sort of question well because they had learnt the rules for simplifying expressions involving indices.

6 Find   **a** the rule   **b** the next two terms   **c** the 12th term for this number sequence.

  102   99   96   93   90

7 Write down   **a** the difference between consecutive terms,

                 b the zero term of this arithmetic sequence.

                     -3 2   7   12   17

8 Here are the first four terms of an arithmetic sequence: 204, 192, 180, 168, …

  a Write down, in terms of n, an expression for the nth term of this arithmetic sequence.

  b Use your answer to part **a** to work out the   **i** 13th term   **ii** 99th term.

9 Here are the first four terms of an arithmetic sequence.

  5   8   11   14

  Is 140 a term in the sequence? You must give a reason for your answer.

D

C

A03

10 Neal is asked to produce an advertising stand for a new variety of soup.
He stacks the cans according to the pattern shown.

The stack is 4 cans high and consists of 10 cans.

a How many cans will there be in a stack 10 cans high?

b Verify that the total number of cans (N) can be calculated by the formula
$N = \dfrac{h(h + 1)}{2}$ when h = number of cans high.

c If he has 200 cans, how high can he make his stack?

11 Naismith, an early Scottish mountain climber, devised a formula that is still used today to calculate
how long it will take mountaineers to climb a mountain. The metric version states:

Allow one hour for every 5 km you walk forward and add on $\frac{1}{2}$ hour for every 300 m of ascent.

a How long should it take to walk 20 km with 900 m of ascent?

A mountain walker's guide contains the following information for a particular walk.

> **Helvellyn Horseshoe**
> Glenridding to Helvellyn via the edges (circular walk)
> Length: 8.5 km
> Total ascent: 800 m
> Time: 4 hour round trip

b Calculate how long this walk should take according to Naismith's formula. Give your answer to the
nearest minute.

c Suggest reasons why this time is different to the one in the guidebook.

12 Simplify

a $(a^5)^4$ b $(3b^4)^2$ c $(3e^5f)^3$

13 The nth even number is $2n$.

Show algebraically that the sum of three consecutive even numbers is always a multiple of 6.

Nov 2008, adapted

14 The expression $\dfrac{6x^2 y}{4y^3}$ can never take a negative value. Explain why.

15 a Simplify $\left(\dfrac{9p^4}{4y^2}\right)^{\frac{1}{2}}$ b Simplify $(2q^3)^{-2}$ c Simplify $\left(\dfrac{12xy^3}{3x^5y}\right)^{\frac{1}{2}}$

16 A 4 by 4 by 4 cube is placed into a tin of yellow paint.

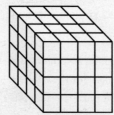

When it has dried, the 64 individual cubes are examined.

How many are covered in yellow paint on 0 sides, 1 side, 2 sides, 3 sides?

Extension: Repeat the question for an n by n by n cube, and show that your expressions add up to n^3.

8 EXPANDING BRACKETS AND FACTORISING

Algebra is regularly used by Formula One teams to maximise the performance of their cars when racing. For example, new rules introduced into Formula One in 2009 have given teams a booster button which gives the car extra power and can be pushed for a maximum of 6.7 seconds during a race. In order to maximise the benefit of the 'boost', F1 teams use algebra to work out the best moment for the driver to use it.

◎ Objectives

In this chapter you will:
- ◎ expand brackets
- ◎ factorise algebraic expressions
- ◎ simplify complicated algebraic expressions.

◐ Before you start

You should be able to:
- ◎ simplify algebraic expressions by collecting like terms
- ◎ use the index law $x^m \times x^n = x^{m+n}$
- ◎ add, subtract, multiply and divide directed numbers.

8.1 Expanding brackets

◎ Objective

○ You can expand expressions which have a single pair of brackets.

⬆ Get Ready

1. Simplify

   a $5 \times 2x$ **b** $3x \times (-4x)$ **c** $(-x) \times (-2x)$

   d $4x - 5 - 3x + 1$ **e** $x^2 + 2x + 3x + 6$ **f** $x^2 - x - 2x + 2$

🔍 Key Points

◎ When there is a number outside a bracket there is a hidden multiplication sign. So $20(n + 3) = 20 \times (n + 3)$.

◎ In algebra, **expand** usually means multiply out.

◎ To expand a bracket you multiply each term inside the bracket by the term outside the bracket.

Example 1   Expand $20(n + 3)$. ← Remember to multiply both terms inside the bracket by 20.

$$20(n + 3) = 20 \times n + 20 \times 3$$
$$= 20n + 60 \quad ← \quad \text{Write your answer in its simplest form.}$$

Example 2   Expand $3(2x + 1)$. ← Multiply both $2x$ and 1 by 3.

$$3(2x + 1) = 3 \times 2x + 3 \times 1$$
$$= 6x + 3$$

Example 3   Expand $p(p + q - 5)$.

$$p(p + q - 5) = p \times p + p \times q - p \times 5 \quad ← \quad p \times 5 \text{ is usually written as } 5p.$$
$$= p^2 + pq - 5p$$

Example 4   Expand $-2x(3x + 1)$. ← Multiply both terms by $-2x$.

$$-2x(3x + 1) = -2x \times 3x + -2x \times 1 \quad ← \quad \text{For each term, negative} \times \text{positive} = \text{negative.}$$
$$= -6x^2 - 2x$$

⚙ Exercise 8A

Questions in this chapter are targeted at the grades indicated.

D

1   Expand

   a $2(x + 3)$ **b** $3(p - 2)$ **c** $4(m + n)$ **d** $3(5 - q)$

   e $2(2x + y - 3)$ **f** $5(2c + 1)$ **g** $4(x^2 - 2)$ **h** $3(n^2 - 2n + 1)$

2 Expand

   a $y(y + 2)$ **b** $g(g - 3)$ **c** $2x(x + 5)$ **d** $n(4 - n)$

   e $a(b + c)$ **f** $s(3s - 4)$ **g** $3t(2t + 1)$ **h** $4x^2(x - 3)$

3 Expand

   a $-2(m + 3)$ **b** $-3(2x + 2)$ **c** $-m(m + 5)$ **d** $-4y(2y + 3)$

   e $-5(p - 2)$ **f** $-3q(1 - q)$ **g** $-2s(s - 3)$ **h** $-3n(4m + n - 5)$

Example 5   Expand and simplify $3(2a + 1) + 2(3a + 5)$.

$$3(2a + 1) + 2(3a + 5) = 6a + 3 + 6a + 10$$

> Expand each bracket separately.

$$= 12a + 13$$

> Collect like terms.

Example 6   Expand and simplify $3x(y - 2) - 2y(x - 3)$.

$$3x(y - 2) - 2y(x - 3) = 3xy - 6x - 2xy + 6y$$
$$= xy - 6x + 6y$$

> For the last term,
> negative × negative = positive.

Example 7   Expand and simplify $6p + 3p(2p - 7) + 4$.

$$6p + 3p(2p - 7) + 4 = 6p + 6p^2 - 21p + 4$$
$$= 6p^2 - 15p + 4$$

ResultsPlus

Watch Out!

You must multiply out the brackets before you collect like terms.
Check your signs.

Exercise 8B

1 Expand and simplify

   a $3(t - 1) + 5t$ **b** $6p + 3(p + 2)$ **c** $6(w + 1) + 5w$

   d $3(d + 2) + 4(d - 2)$ **e** $3a + b + 2(a + b)$ **f** $2(5x - y) + 5(y - x + 1)$

2 Expand and simplify

   a $3(y + 10) - 2(y + 5)$ **b** $6(2a + 1) - 3(a + 4)$ **c** $x - 5(x + 3)$

   d $q(q + 3) - 3(q + 1)$ **e** $2n(n - 2) - n(2n + 1)$ **f** $3m(2 + 5m) - 4m(1 + m)$

3 Expand and simplify

   a $5(t - 4) - 4(t - 1)$ **b** $3(x + 3) - 2(x - 5)$ **c** $2g(g + 1) - g(g + 1)$

   d $6c(2c - 3) - c(4 - c)$ **e** $4s(s + 3) - 2(1 - s)$ **f** $p(p + q) - q(p - q)$

4 Expand and simplify

   a $7s - 4(s + 1)$ **b** $12m + 3(m + 2)$ **c** $8f^2 - 3f(f + 1)$

   d $5n + n(n - 1)$ **e** $2x - x(x - y)$ **f** $7p - 2p(1 - p)$

8.2 Factorising by taking out common factors

◎ Objectives

- ○ You can recognise common factors.
- ○ You can factorise expressions by taking out common factors.

⬆ Get Ready

1. What is the HCF (Highest Common Factor) of the following pairs?
   a 6 and 8   **b** 10 and 25   **c** $8x$ and 12   **d** $9y$ and $15y$

⬤ Key Points

- ◉ **Factorising** is the opposite of expanding brackets, as you will need to put brackets in.
- ◉ To factorise an expression, find a common factor of the terms, take this factor outside the brackets, then decide what is needed inside the brackets.
- ◉ You can check your answer by expanding the brackets.
- ◉ Common factors are not always single terms such as 2, $5x$, $3a^2b$.
- ◉ Sometimes a common factor can have more than one term, for example $x + 2$ or $2a - b$.

Example 8   Factorise $12b + 8$

$$12b + 8 = 4(\quad)$$
$$= 4(3b + 2)$$

> The common factor of $12b$ and 8 is 4.
> Note that you would not usually write the 4( ) but it is there to remind you to find the common factor first.

> Check this multiplies out to give $12b + 8$.

Example 9   Factorise $2 - 6y$

$$2 - 6y = 2(\quad)$$
$$= 2(1 - 3y)$$

> Pick out the common factor first.
> 1 is needed as the first term in the bracket.

Example 10   Factorise $x^2 + 3x$

$$x^2 + 3x = x(\quad)$$
$$= x(x + 3)$$

> The common factor of x^2 and $3x$ is x.
> Remember to check by multiplying out.

Example 11   Factorise $15p - 10q - 20pq$

$$15p - 10q - 20pq = 5(\quad)$$
$$= 5(3p - 2q - 4pq)$$

> Find the common factor of all three terms.

Example 12 ▶ Factorise completely $6a^2b + 9ab^2$

$6a^2b + 9ab^2 = 3ab(\quad)$ ◀─── The common factor of $6 \times a \times a \times b$ and $9 \times a \times b \times b$ is $3 \times a \times b$.
$\qquad\qquad\;\; = 3ab(2a + 3b)$

Exercise 8C

1 Factorise

a $3x + 6$ b $2y - 2$ c $5p + 10q$ d $14t - 7$

e $8s + 2t$ f $9a + 18b$ g $15u + 5v + 10w$ h $xt - yt$

i $ac - c$ j $6x^2 + 9x + 3$ k $2p^2 - 2p$ l $q^2 - q$

m $4x^2 + 3x$ n $2h - 5h^2$ o $p^3 + 2p$ p $s^2 + s^3$

2 Factorise completely

a $5xy + 5xt$ b $3ad - 6ac$ c $6pq + 4hp$ d $8xy - 4y$

e $4pq + 2ps + 8pt$ f $mn - kmn$ g $2x^2 + 4x$ h $12s^2 - 24s$

i $6f^2 + 2f^3$ j $y^4 + y^2$ k $3cd^2 - 5c^2d$ l $a^3b + ab^3$

m $8pqr + 10prs$ n $14a^2b - 7ab^2 + 21ab$ o $15x^2y - 35x^2y^2$ p $(3y)^2 + 3y$

Example 13 ▶ Factorise $5(x + 2)^2 - 3(x + 2)$

$5(x + 2)^2 - 3(x + 2) = (x + 2)[\quad]$ ◀─── $(x + 2)$ is a common factor.
$\qquad\qquad\qquad\quad\; = (x + 2)[5(x + 2) - 3]$
$\qquad\qquad\qquad\quad\; = (x + 2)[5x + 10 - 3]$ ◀─── Simplify the expression inside the square bracket.
$\qquad\qquad\qquad\quad\; = (x + 2)(5x + 7)$

Example 14 ▶ Factorise completely $12(s + 2t) - 4(s + 2t)^2$

$12(s + 2t) - 4(s + 2t)^2 = 4(s + 2t)[\quad]$ ◀─── The common factor is $4(s + 2t)$.
$\qquad\qquad\qquad\qquad\quad = 4(s + 2t)[3 - (s + 2t)]$ ◀─── This cannot be simplified further.
$\qquad\qquad\qquad\qquad\quad = 4(s + 2t)(3 - s - 2t)$

Exercise 8D

1 Factorise

a $(x + 3)^2 + 2(x + 3)$ b $x(x - y) + y(x - y)$ c $p(p + 4) - 3p$

d $(2t + s)(2t - s) + (2t - s)$ e $(a - 5)^2 - 2(a - 5)$ f $(2d + 1)^2 + (2d + 1)$

2 Factorise completely

a $2(y + 2)^2 + 4(y + 2)$ b $15(x - 1)^2 - 10(x - 1)$ c $8(p + 5)^2 + 10(p + 5)$

d $9(q + 1) + 6(q + 1)^2$ e $7(a - b)(a + b) - 14(a + b)$ f $4x^2(x + 1) - 6x(x + 1)$

D

C

B

83

8.3 Expanding the product of two brackets

○ You can multiply out the product of two brackets.

Get Ready

1. Work out the area of this rectangle.

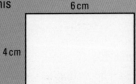

2. Write down an expression, in terms of x, for the area of this rectangle.

Key Points

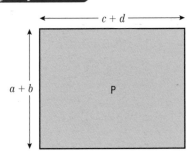

Area of rectangle P $= (a + b) \times (c + d)$
$= (a + b)(c + d)$

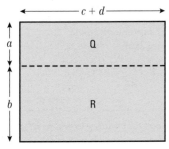

Area of rectangle Q $= a(c + d)$
Area of rectangle R $= b(c + d)$
Area of rectangle P $=$ Area of rectangle Q $+$ Area of rectangle R
$(a + b)(c + d) \quad = a(c + d) + b(c + d)$
$= ac + ad + bc + bd$

⊙ To multiply out the product of two brackets:
  ⊙ multiply each term in the first bracket by the second bracket
  ⊙ expand the brackets
  ⊙ simplify the resulting expression.
⊙ An alternative method is the **grid method** (see Example 15).

Example 15   Expand and simplify $(x + 2)(x + 3)$.

Method 1
$(x + 2)(x + 3) = x(x + 3) + 2(x + 3)$
$= x^2 + 3x + 2x + 6$
$= x^2 + 5x + 6$

> Take each term in the first bracket, in turn, and multiply it by the second bracket.
> Expand the brackets.
> Collect the like terms.

Method 2 — the grid method

	x	$+3$
x	x^2	$+3x$
$+2$	$+2x$	$+6$

Each term in the first bracket is multiplied by each term in the second bracket.

$(x + 2)(x + 3) = x^2 + 3x + 2x + 6$

Add the four terms highlighted.

$= x^2 + 5x + 6$

Collect like terms.

Example 16   Expand and simplify $(m + 2)^2$

$$(m + 2)^2 = (m + 2)(m + 2)$$
$$= m(m + 2) + 2(m + 2)$$
$$= m^2 + 2m + 2m + 4$$
$$= m^2 + 4m + 4$$

Write out $(m + 2)^2$ in full.

Watch Out!

Note that $(a + b)^2$ is not equal to $a^2 + b^2$.

Example 17   Expand and simplify $(2t - 1)(3t - 2)$

Method 1

$$(2t - 1)(3t - 2) = 2t(3t - 2) - 1(3t - 2)$$
$$= 6t^2 - 4t - 3t + 2$$
$$= 6t^2 - 7t + 2$$

Check your signs are correct.

Method 2

	$3t$	-2
$2t$	$6t^2$	$-4t$
-1	$-3t$	$+2$

$$(2t - 1)(3t - 2) = 6t^2 - 4t - 3t + 2$$
$$= 6t^2 - 7t + 2$$

Exercise 8E

1   Expand and simplify

   a $(x + 3)(x + 4)$ **b** $(x + 1)(x + 2)$ **c** $(x + 2)(x - 5)$

   d $(y - 2)(y + 3)$ **e** $(y + 1)(y - 2)$ **f** $(x - 2)(x - 3)$

   g $(a - 4)(a - 5)$ **h** $(x + 2)^2$ **i** $(p + 4)^2$

   j $(k - 7)^2$ **k** $(a + b)^2$ **l** $(a - b)^2$

B

A

2 Expand and simplify

a $(x + 1)(2x + 1)$

b $(x - 1)(3x + 1)$

c $(2x + 3)(x + 4)$

d $(y - 3)(3y + 1)$

e $(2p + 1)(p + 3)$

f $(2t + 1)(3t + 2)$

g $(3s + 2)(2s + 5)$

h $(2x - 3)(2x + 5)$

i $(3y + 2)(4y - 1)$

j $(2a - 1)(3a - 2)$

k $(3x + 2)^2$

l $(2k - 1)^2$

3 Expand and simplify

a $(x + y)(x + 2y)$

b $(x - y)(x + 2y)$

c $(x + y)(x - 2y)$

d $(x - y)(x - 2y)$

e $(2p + 3q)(3p - q)$

f $(3s - 2t)(2s - t)$

g $(2a + 3b)^2$

h $(2a - 3b)^2$

8.4 Factorising quadratic expressions

◎ Objectives

- You can factorise quadratic expressions of the form $x^2 + bx + c$.
- You can recognise and factorise the difference of two squares.
- You can factorise quadratic expressions of the form $ax^2 + bx + c$.

⬆ Get Ready

1. Write down all possible pairs of numbers whose product is

   a -6 b 15.

2. Find a pair of numbers whose product is 10 and whose sum is 7.

3. Find a pair of numbers whose product is 15 and whose sum is -8.

Key Points

- Factorising is the reverse process to expanding brackets so, for example, factorising $x^2 + 5x + 6$ gives $(x + 2)(x + 3)$.

- To factorise the quadratic expression $x^2 + bx + c$
  - find two numbers whose product is $+c$ and whose sum is $+b$
  - use these two numbers, p and q, to write down the factorised form $(x + p)(x + q)$.

- To factorise the quadratic expression $ax^2 + bx + c$
  - work out the value of ac
  - find a pair of numbers whose product is $+ac$ and sum is $+b$
  - rewrite the x term in the expression using these two numbers
  - factorise the first two terms and the last two terms
  - pick out the common factor and write as the product of two brackets.

- Any expression which may be written in the form $a^2 - b^2$, known as the difference of two squares, can be factorised using the result $a^2 - b^2 = (a + b)(a - b)$.

Example 18 Factorise $x^2 + 7x + 12$

The pairs of numbers whose product is 12 are:

$$+1 \times +12 \qquad -1 \times -12$$
$$+2 \times +6 \qquad -2 \times -6$$
$$+3 \times +4 \qquad -3 \times -4$$

$$+3 \times +4 = +12 \leftarrow$$
$$+3 + +4 = +7$$
$$x^2 + 7x + 12 = (x + 3)(x + 4) \leftarrow$$

Find two numbers whose product is $+12$ and whose sum is $+7$.

Put into factorised form using the numbers $+3$ and $+4$.

ResultsPlus
Examiner's Tip

You may find it helpful to start by writing down all the pairs of numbers whose product is $+12$.

Example 19 Factorise $x^2 - 10x + 25$

The pairs of numbers whose product is $+25$ are:

$$+1 \times +25 \qquad -1 \times -25$$
$$+5 \times +5 \qquad -5 \times -5$$

$$-5 \times -5 = +25$$
$$-5 + -5 = -10$$
$$x^2 - 10x + 25 = (x - 5)(x - 5) \leftarrow$$

This may also be written as $(x - 5)^2$.

ResultsPlus
Examiner's Tip

You can check your answer by expanding the brackets.

Exercise 8F

1 Write down a pair of numbers:
   a whose product is $+15$ and whose sum is $+8$
   b whose product is $+24$ and whose sum is -10
   c whose product is $+18$ and whose sum is -9
   d whose product is -8 and whose sum is $+2$
   e whose product is -8 and whose sum is -2
   f whose product is -9 and whose sum is 0.

2 Factorise
   a $x^2 + 8x + 15$ b $x^2 + 8x + 7$ c $x^2 + 9x + 20$
   d $x^2 + 6x + 9$ e $x^2 - 6x + 5$ f $x^2 - 2x + 1$
   g $x^2 + 3x - 18$ h $x^2 - 3x - 18$ i $x^2 + 3x - 28$
   j $x^2 - x - 12$ k $x^2 + 2x - 24$ l $x^2 - 4$
   m $x^2 - 81$

A

Example 20 Factorise $x^2 - n^2$

$$x^2 - n^2 = x^2 + 0x - n^2$$

The pair of numbers whose product is $-n^2$ and whose sum is 0 is $+n \times -n$:

$$x^2 - n^2 = (x + n)(x - n)$$

Example 21 Factorise $x^2 - 100$

$$x^2 - 100 \quad = x^2 - 10^2$$
$$= (x + 10)(x - 10)$$

> Substitute $a = x$ and $b = 10$ into $a^2 - b^2 = (a + b)(a - b)$.

Results Plus
Examiner's Tip

It will help you in the examination if you learn $a^2 - b^2 = (a + b)(a - b)$.

Example 22

a Factorise $p^2 - q^2$

b Hence, without using a calculator, find the value of $101^2 - 99^2$

a $p^2 - q^2 = (p + q)(p - q)$

b $101^2 - 99^2$

> Use the result $a^2 - b^2 = (a + b)(a - b)$.
> Substitute $p = 101$ and $q = 99$ in the answer to part (a).

$= (101 + 99)(101 - 99)$

$= 200 \times 2$

> Work out each bracket.

$= 400$

Example 23 Factorise $(x + y)^2 - 4(x - y)^2$

$$= (x + y)^2 - [2(x - y)]^2$$

> Write $(x + y)^2 - 4(x - y)^2$ in the form $a^2 - b^2$.

$$= [(x + y) + 2(x - y)][(x + y) - 2(x - y)]$$

> Substitute $a = (x + y)$ and $b = 2(x - y)$ into $a^2 - b^2 = (a + b)(a - b)$.

$$= [x + y + 2x - 2y][x + y - 2x + 2y]$$

> Expand and simplify the expression in each square bracket.

$$(x + y)^2 - 4(x - y)^2 = (3x - y)(-x + 3y)$$

> Note that alternatively this answer may be written as $(3x - y)(3y - x)$.

Exercise 8G

A

1 Factorise

   a $x^2 - 36$ b $x^2 - 49$ c $y^2 - 144$

   d $25 - y^2$ e $w^2 - 2500$ f $10\,000 - a^2$

   g $(x + 1)^2 - 4$ h $81 - (9 - y)^2$ i $(a + b)^2 - (a - b)^2$

2 Without using a calculator, find the value of:

   a $64^2 - 36^2$ b $7.5^2 - 2.5^2$ c $0.875^2 - 0.125^2$ d $1005^2 - 995^2$

3 Factorise these expressions, simplifying your answers where possible.

   a $4x^2 - 49$ b $9y^2 - 1$ c $121t^2 - 400$

   d $1 - (q + 2)^2$ e $(2t + 1)^2 - (2t - 1)^2$ f $(p + q + 1)^2 - (p + q - 1)^2$

   g $100(p + \frac{1}{2})^2 - 4(q + \frac{1}{2})^2$ h $25(s + t)^2 - 25(s - t)^2$

A

4 Factorise completely

a $3x^2 - 12$ b $5y^2 - 125$ c $10w^2 - 1000$

d $4p^2 - 64q^2$ e $12a^2 - 27b^2$ f $2(x + 1)^2 - 2(x - 1)^2$

Example 24   Factorise $3x^2 - 7x + 4$.

$a = +3, b = -7, c = +4$ ← Find two numbers whose product is $+12$ and whose sum is -7.

$ac = 12, b = -7$

$-3 \times -4 = +12$ ← Replace $-7x$ with $-3x - 4x$.

$-3 + -4 = -7$

$3x^2 - 7x + 4 = 3x^2 - 3x - 4x + 4$ ← Factorise by grouping.

$= 3x(x - 1) - 4(x - 1)$ ← Pick out the common factor and write as the product of two brackets.

$= (x - 1)(3x - 4)$

$3x^2 - 7x + 4 = (x - 1)(3x - 4)$

Exercise 8H

A

1 Factorise

a $5x^2 + 16x + 3$ b $2x^2 + 11x + 5$ c $3x^2 + 4x + 1$ d $8x^2 + 6x + 1$

e $6x^2 + 13x + 6$ f $6x^2 - 7x + 1$ g $5x^2 - 7x + 2$ h $12x^2 - 11x + 2$

i $8x^2 + 2x - 3$ j $2x^2 - 7x - 15$ k $7x^2 - 19x - 6$ l $3x^2 - 10x - 8$

m $4y^2 + 12y + 5$ n $6y^2 - 13y + 2$ o $6y^2 - 25y + 25$

2 Factorise completely

a $6x^2 + 14x + 8$ b $6y^2 - 15y + 6$ c $5x^2 + 5x - 10$

3 Factorise

A*

a $x^2 + xy - 2y^2$ b $2x^2 + 7xy + 5y^2$ c $6x^2 + 5xy - 6y^2$

Chapter review

⦿ When there is a number outside a bracket there is a hidden multiplication sign.

⦿ In algebra, **expand** usually means multiply out.

⦿ To expand a bracket you multiply each term inside the bracket by the term outside the bracket.

⦿ **Factorising** is the opposite of expanding brackets, as you will need to put brackets in.

⦿ To factorise an expression, find the common factor of the terms, take this factor outside the brackets, decide what is needed inside the brackets.

⦿ You can check your answer by expanding the brackets.

⦿ Common factors are not always single terms.

- To multiply out the product of two brackets:
  - multiply each term in the first bracket by the second bracket
  - expand the brackets
  - simplify the resulting expression.
- An alternative method is the **grid method**.
- Factorising is the reverse process to expanding brackets.
- To factorise the quadratic expression $x^2 + bx + c$
  - find two numbers whose product is $+c$ and whose sum is $+b$
  - use these two numbers, p and q, to write down the factorised form $(x + p)(x + q)$.
- To factorise the quadratic expression $ax^2 + bx + c$
  - work out the value of ac
  - find a pair of numbers whose product is $+ac$ and sum is $+b$
  - rewrite the x term in the expression using these two numbers
  - factorise the first two terms and the last two terms
  - pick out the common factor and write as the product of two brackets.
- Any expression which may be written in the form $a^2 - b^2$, known as the difference of two squares, can be factorised using the result $a^2 - b^2 = (a + b)(a - b)$.

Review exercise

1   Expand and simplify $2(x - 4) + 3(x + 2)$

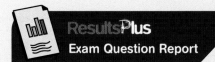

87% of students answered this sort of question well because they remembered to do all of the necessary multiplications.

June 2009

2   **a** Factorise $5m + 10$ **b** Factorise $y^2 - 3y$

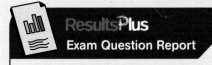

50% of students answered this question poorly because they did not put factors in the right place.

Nov 2008

3   Factorise **a** $ax + by + bx + ay$ **b** Factorise $ac - bd + ad - bc$

4   Expand and simplify $(x + 4)(x - 3)$ *June 2009*

5   Expand **a** $(a + 2)^2$ **b** $(c - 3)^2$ **c** $(d + 1)^2$ **d** $(x + y)^2$

6   Expand and simplify
   a $(x + 5)(x + 10)$ **b** $(y + 9)^2$ **c** $(x - 4)(x + 2)$ **d** $(x + 2)(x - 3)$ **e** $(t - 1)(t - 6)$
   f $(x + 4)(2x + 3)$ **g** $(3p - 1)(2p + 1)$ **h** $(2c + d)(2c - d)$ **i** $(4y - 1)^2$

7 Factorise

a $t^2 + 11t + 30$ b $x^2 + 14x + 49$ c $p^2 + 2p - 15$

d $y^2 - 12y + 36$ e $x^2 - 5x + 4$ f $s^2 - 64$

8 a Factorise $x^2 + 8x + 7$ b Express 187 as the product of 2 prime numbers.

9 Jamie is planting flowers in a local park.

$$R \quad R \quad R \quad R \quad R$$
$$R \quad Y \quad Y \quad Y \quad R$$
$$R \quad R \quad R \quad R \quad R$$

When he plants three yellow flowers, he surrounds them with twelve red flowers, as shown in the diagram.

a How many red flowers (R) does he plant with ten yellow flowers (Y)?

b How many red flowers would he plant with y yellow flowers ?

Write your answer in both factorised and unfactorised forms.

10 Factorise

a $x^2 - 400$ b $9t^2 - 4$ c $100 - y^2$ d $25 - 4p^2$

11 Use some of your answers to question 10 to work out the values of the following expressions without using a calculator.

a $21^2 - 20^2$ b $10^2 - 9.9^2$ c $5^2 - 3^2$

12 Work out $1002^2 - 998^2$ using algebra.

13 For any three consecutive numbers show that the difference between the product of the first and second and the product of the second and third is equal to double the second number.

14 Factorise fully $3(x + 2)^2 - 3x(x + 2)$

15 a In the group stage of the Champions League, four teams play each other both at home and away. Prove that this requires 12 matches in total.

b Similarly, in the Premier League all the teams play each other twice. There are 20 teams. How many games are there altogether?

c How many games are there in a league with a teams? Write your answer in a factorised and unfactorised form.

16 Factorise

a $2x^2 + 5x + 2$ b $2w^2 + 5w - 3$ c $3a^2 + 14a + 8$

d $30z^2 - 23z + 2$ e $8y^2 + 23y - 3$ f $6p^2 - pq - q^2$

17

1	2	3	4	5	6
7	8	9	10	11	12
13	14	15	16	17	18

$3 \times 8 - 2 \times 9 = 6$ and $12 \times 17 - 11 \times 18 = 6$

Show that for any 2 by 2 square of numbers from the grid, the difference of the products of numbers from opposite corners is always 6.

B

A02
A03

A

A02
A03

A03

A03

A*

A03

9 GRAPHS

The street system of New York is made up of straight lines running at right angles to each other. The city did not have this layout from the start, but in 1811 the mesh of rectangles which characterises Manhattan today was built over the original tangle of city streets. Lanes and paths that didn't fit into the pattern were blocked up and the buildings that lined them torn down. The only street that survived was Broadway – an original Indian track which angled across the island and can still be seen in the city.

◉ Objectives

In this chapter you will:
- draw straight-line graphs
- find the midpoint of a line segment
- find the gradient, y-intercept and equation of a straight line
- understand the relationship between the gradients of parallel and perpendicular lines
- draw and interpret graphs describing real-life situations, including distance-time graphs
- consider compound measures and solve problems involving average speed.

◐ Before you start

You need to be able to:
- draw, label and scale axes
- substitute numbers in simple algebraic expressions
- solve simple equations
- draw straight lines by plotting points.

9.1 Drawing straight-line graphs by plotting points

◎ Objectives

○ You can use the equation of a straight line to find the coordinates of points on the line.
○ You can draw graphs of straight lines.

⊘ Why do this?

Straight lines are often used to show the relationship between quantities, for example, to convert money between pounds and euros.

⬥ Get Ready

1. Work out the value of $2x + 3$ if
  a $x = 2$ **b** $x = -1$

2. Solve the equations
  a $2 + y = 3$ **b** $3 + 2y = 6$

Key Points

◉ To draw a straight line by plotting points you need to plot at least two points which fit the equation of the line.
◉ The **equation of a straight line** can have several forms:
  ◎ lines in the form $x = c$ or $y = c$ where c is a number (see Example 1 and Exercise 9A)
  ◎ lines in the form $y = mx + c$ where m and c are numbers (see Examples 2 and 3 and Exercise 9B)
  ◎ lines in the form $ax + by = c$ where a, b and c are numbers (see Example 4 and Exercise 9C).

Example 1   Draw the graph of $x = 2$.

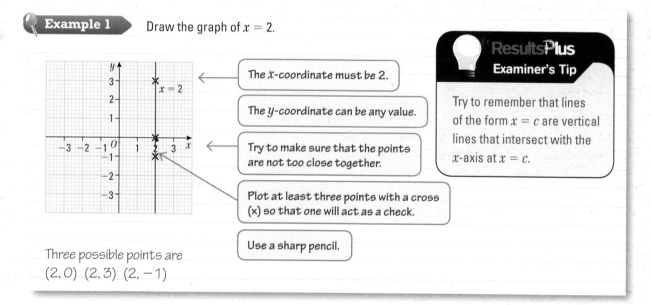

The x-coordinate must be 2.

The y-coordinate can be any value.

Try to make sure that the points are not too close together.

Plot at least three points with a cross (x) so that one will act as a check.

Use a sharp pencil.

ResultsPlus
Examiner's Tip

Try to remember that lines of the form $x = c$ are vertical lines that intersect with the x-axis at $x = c$.

Three possible points are
$(2, 0)$ $(2, 3)$ $(2, -1)$

⚙ Exercise 9A

Questions in this chapter are targeted at the grades indicated.

1 Draw, on separate axes, the graphs of:
  a $x = -3$ **b** $y = 2$ **c** $x = 0$
  d $y = 0$ **e** $x = 1.5$ **f** $y = -5$

2 Write down the equation of each of these lines.

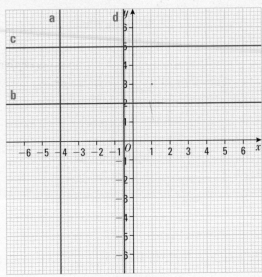

3 Write down the coordinates of the point where the following pairs of lines cross.

    a $x = 1, y = 3$ **b** $x = -4, y = 2$ **c** $y = 3, x = -\frac{1}{2}$

4 Find the perimeter and area of the rectangle formed by the lines $x = -3$, $x = 1$, $y = -2$ and $y = 5$.

D   A02   A03

Example 2   Draw the graphs of $y = x$ and $y = -x$.

$y = x$

when $x = 0$, $y = 0$ $(0, 0)$

when $x = 3$, $y = 3$ $(3, 3)$

when $x = -3$, $y = -3$ $(-3, -3)$

> Find the coordinates of at least two points for which the y value is the same as the x value.

$y = -x$

when $x = 0$, $y = 0$ $(0, 0)$

when $x = 3$, $y = -3$ $(3, -3)$

when $x = -3$, $y = 3$ $(-3, 3)$

> Find the coordinates of at least two points for which the y value is the negative of the x value – that is, it has the opposite sign.

Results Plus

Examiner's Tip

These diagonal lines occur frequently so it is useful to be able to draw them from memory.

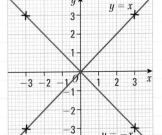

> Plot the three points for each graph and join them with a ruler.

> Label each of the lines with their equation.

Example 3　　Draw the graph of $y = 3 - 2x$.
　　　　　　　　Use values of x from -2 to $+5$.

When $x = 0$,　　$y = 3 - 2 \times 0 = 3$　　$(0, 3)$　　← Use $x = 0$ to make
When $x = 5$,　　$y = 3 - 2 \times 5 = -7$　　$(5, -7)$　　the working out
When $x = -2$,　$y = 3 - 2 \times (-2) = 7$　$(-2, 7)$　　easier.

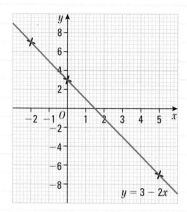

$y = 3 - 2x$

Exercise 9B

1　a　Copy and complete the table of values for $y = 4 - 2x$.

x	-2	-1	0	1	2	3	4
y			4			-2	-4

　b　Draw the graph of $y = 4 - 2x$. Use values of x from -2 to $+4$.

2　a　Draw the graph of $y = 4x - 8$. Use values of x from -1 to $+4$.
　b　Write down the coordinates of the point where the graph intersects:
　　i　the x-axis　　　　　　　　ii　the y-axis.
　c　Use your graph to find:
　　i　the value of y when $x = 2.5$　　ii　the value of x when $y = 6$.

3　a　On the same axes, draw the graphs of $y = x$, $y = -x$, $y = 2x$ and $y = -3x$.
　b　What can you say about the graphs of all lines with equations of the form $y = mx + c$ with $c = 0$?

4　a　On the same axes, draw the graphs of $y = 6x - 9$ and $y = 9 - 6x$.
　b　Write down the coordinates of the point where the two graphs intersect.

5　Work out the area of the triangle formed by the lines $y = 0$, $y = x$ and $y = 2x + 5$.

D

A03

A02
A03

C

Example 4 Draw the graph of $2x + y = 6$.

When $x = 0, 2 \times 0 + y = 6$ $y = 6$ $(0, 6)$ ← Use $x = 0$ to find one point.

When $y = 0, 2x + 0 = 6$ $x = 3$ $(3, 0)$ ← Use $y = 0$ to find a second point. Solve $2x = 6$.

When $x = 2, 2 \times 2 + y = 6$ $4 + y = 6$ $y = 2$ $(2, 2)$ ← Choose a value of x to use as a check point and work out y. Solve $4 + y = 6$.

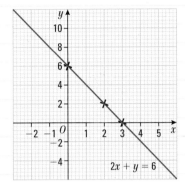

Examiner's Tip

When you have equations in the form $ax + by = c$, it is easier if you use $x = 0$ and $y = 0$ to find two of your points.

Exercise 9C

C

1 Draw the graph of $3x + 4y = 12$.

A03

2 **a** On the same axes, draw the graphs of:
   i $x + y = 2$ **ii** $x + y = 7$ **iii** $x + y = -4$.
   b What do you notice about the graphs you have drawn?

A02

3 Find the coordinates of the point where the lines $2x + 3y = 12$ and $x - y = 1$ intersect.

A02
A03

4 **a** On the same axes, draw the graphs of $5x + 2y = 10$ and $5y - 2x = 8$.
   b What do you notice about the way in which these two graphs intersect?
   c Write down the equations of two other lines that intersect in this way.

9.2 Finding the midpoint of a line segment

⊙ **Objective**

○ You can find the coordinates of the midpoint of a line segment.

❓ **Why do this?**

You might do this if you and a friend agree to meet halfway between your house and theirs.

⬆ **Get Ready**

1. Find the number halfway between:
   a 4 and 6   **b** 5 and 6   **c** 17 and 17.5   **d** 0.20 and 0.25
   e −2 and 4   **f** −6 and −11

Key Points

● A line joining two points is called a line segment.
AB is the line segment joining points A and B.

● The **midpoint** of a line is halfway along the line.

● To find the midpoint you should add the x-coordinates and divide by 2, and add the y-coordinates and divide by 2.

● The midpoint of the line segment AB between
A (x_1, y_1) and B (x_2, y_2) is $\left(\dfrac{x_1 + x_2}{2}, \dfrac{y_1 + y_2}{2}\right)$.

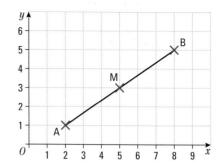

Example 5   Work out the coordinates of the midpoint of the line segment PQ where P is (2, 3) and Q is (7, 11).

x-coordinate $2 + 7 = 9$ ←   [Add the x-coordinates and divide by 2.]
$9 \div 2 = 4\frac{1}{2}$

y-coordinate $3 + 11 = 14$ ←   [Add the y-coordinates and divide by 2.]
$14 \div 2 = 7$

The midpoint is $(4\frac{1}{2}, 7)$.

Example 6   Find the midpoint of RS.

R has coordinates $(-3, 1)$
S has coordinates $(3, 4)$

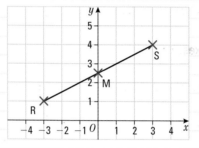

Add the x-coordinates and divide by 2.
$-3 + 3 = 0$ $0 \div 2 = 0$
Add the y-coordinates and divide by 2.
$1 + 4 = 5$ $5 \div 2 = 2.5$
M has coordinates $(0, 2.5)$ or $(0, 2\frac{1}{2})$.

Exercise 9D

1   Work out the coordinates of the midpoint of each of the line segments shown on the grid.

   a OA   **b** BC   **c** DE

   d FG   **e** HJ   **f** KL

   g MN   **h** PQ   **i** ST

   j UV

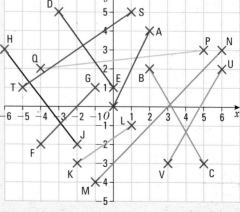

D

D

2 Work out the coordinates of the midpoint of each of these line segments.

a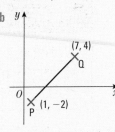

b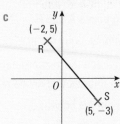

c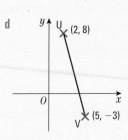

d

C

3 Work out the coordinates of the midpoint of each of these line segments.

a   AB when A is $(-1, -1)$ and B is $(9, 9)$ b   PQ when P is $(2, -4)$ and Q is $(-6, 9)$

c   ST when S is $(5, -8)$ and T is $(-2, 1)$ d   CD when C is $(1, 7)$ and D is $(-7, 2)$

e   UV when U is $(-2, 3)$ and V is $(6, -8)$ f   GH when G is $(-2, -6)$ and H is $(7, 3)$

9.3 The gradient and y-intercept of a straight line

⊙ Objectives

○ You can find the gradient of a straight line from its graph.
○ You can find the y-intercept of a straight line from its graph.
○ You can interpret the gradient of a straight line.

⊘ Why do this?

Gradients on graphs often represent important measures in real-life such as speed, acceleration or cost of petrol per litre.

⬆ Get Ready

1. Draw, on the same pair of axes, the graphs of $y = x$, $y = 2x$, $y = 3x$
2. Draw, on the same pair of axes, the graphs of $y = -x$, $y = -\frac{1}{2}x$, $y = -2x$
3. Draw, on the same pair of axes, the graphs of $y = 3x - 2$, $y = 3x$, $y = 3x - 3$

⚲ Key Points

● The **gradient** of a straight line is a measure of its slope.
● Steeper lines have larger gradients.
● Gently sloping lines have smaller gradients.
● Gradient of a line $= \dfrac{\text{change in } y\text{-direction}}{\text{change in } x\text{-direction}}$
● Lines which slope upwards from left to right have positive gradients.
● Lines which slope downwards from left to right have negative gradients.
● The gradient of the line through the points (x_1, y_1) and (x_2, y_2) is given by $m = \dfrac{y_2 - y_1}{x_2 - x_1}$

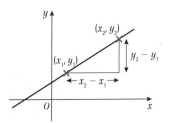

● The **y-intercept** of a line is the value of y when $x = 0$.
  It is shown by the point where the graph crosses the y-axis.

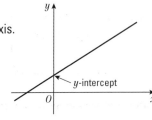

Example 7 Find the gradient and y-intercept of this straight-line graph.

ResultsPlus
Examiner's Tip

Working out the gradient is easier if you choose a large triangle with a base which is a whole number of units.

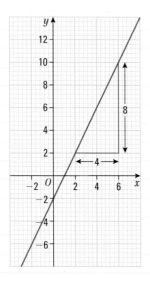

$$\text{Gradient} = \frac{\text{change in } y\text{-direction}}{\text{change in } x\text{-direction}}$$ ← Draw a right-angled triangle on the line.

$$= \frac{10 - 2}{6 - 2}$$ ← Find the difference in the y values and the difference in the x values.

$$= \frac{8}{4}$$ ← Work out the value of the fraction.

Gradient $= 2$

Intercept $= -2$ ← Read off the value where the graph crosses the y-axis.

Example 8 Find the gradient and the coordinates of the y-intercept of this straight-line graph.

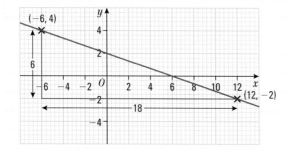

$$\text{Gradient} = \frac{\text{change in } y\text{-direction}}{\text{change in } x\text{-direction}}$$

$$= -\frac{6}{18}$$ ← Put a negative sign in as the gradient slopes down from left to right.

$$= -\frac{1}{3}$$ ← Simplify the fraction if possible.

y-intercept $= 2$
y-intercept has coordinates $(0, 2)$.

ResultsPlus
Examiner's Tip

Make sure you take into account the scales on the axes when finding the lengths of the sides of the triangle.

Example 9   Find the gradient of the line joining the points A $(-5, -1)$ and B $(4, 5)$.

Method 1

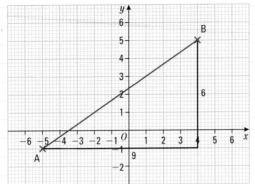

Draw a diagram to help you.

$$\frac{\text{vertical distance}}{\text{horizontal distance}} = \frac{\text{difference in } y\text{-coordinates}}{\text{difference in } x\text{-coordinates}}$$

$$= \frac{5 + 1}{4 + 5}$$

$$= \frac{6}{9}$$

Gradient $= \dfrac{2}{3}$

Method 2

$$m = \frac{y_2 - y_1}{x_2 - x_1}$$

$$= \frac{5 - (-1)}{4 - (-5)}$$

$$= \frac{6}{9}$$

Gradient $= \dfrac{2}{3}$

Use $(x_1, y_1) = (-5, -1)$ and $(x_2, y_2) = (4, 5)$.
Put $x_1 = -5, y_1 = -1$ and $x_2 = 4, y_2 = 5$ into the formula for m.

Exercise 9E

D

1   Work out the gradient of each line.

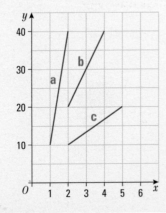

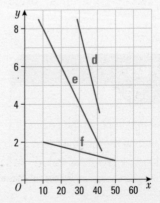

2 Work out the gradient and y-intercept of each straight line.

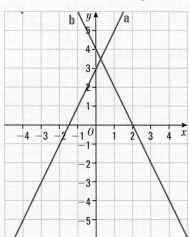
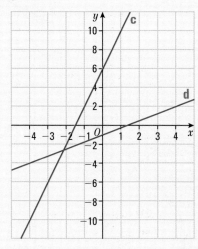

3 Using separate axes, using 1 cm to represent 1 unit, draw straight lines with:
  a gradient 1, y-intercept 2
  b gradient 2, y-intercept -3
  c gradient $\frac{1}{2}$, y-intercept 4
  d gradient -3, y-intercept 3
  e gradient $-\frac{1}{3}$, passing through the point $(0, 2)$.

4 A straight line has gradient 3. The point $(2, -1)$ lies on the line. Find the coordinates of one other point on the line.

5 A is the point $(-4, 6)$. B is the point $(8, 0)$.
  a Find the gradient of the line AB.
  b Find the coordinates of the y-intercept of the line AB.

Example 10 The graph shows the charge for gas, in £s, supplied to a customer by a gas company.
The cost consists of a standing charge plus a charge for each unit of gas used.
  a Write down the standing charge made by the gas company.
  b Work out the gradient of the graph and explain what it represents.

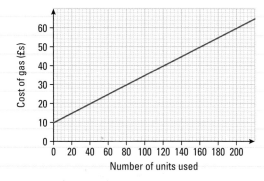

a

Cost of gas (£s) / Number of units used graph

Standing charge

> Draw a triangle on the line and work out the gradient.

> Read off the cost of 0 units on the graph.

Standing charge = £10

b Gradient = $\frac{25}{100}$

= 0.25

> The gradient is found by dividing the cost (in £s) by the number of units used. It represents the cost of each unit of gas used.

The cost of gas is 25p per unit used.

Exercise 9F

D

1 The graph shows the cooking time, t minutes, needed for a chicken of weight w kilograms.

a Work out the gradient of the graph and explain what it represents.

b Describe a rule to give the time needed to cook a chicken of any weight.

c Why do you think the line doesn't extend down to the x-axis?

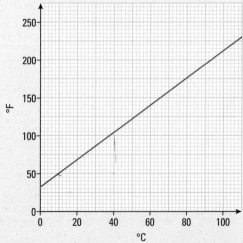

2 This graph can be used to change between temperatures measured in °C and temperatures measured in °F.

a What temperature in °F is equivalent to 10 °C?

b Find the gradient of this graph and explain what it represents.

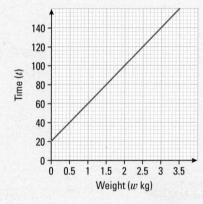

3   The diagram shows the distance–time graphs of a car, a cycle and a lorry.

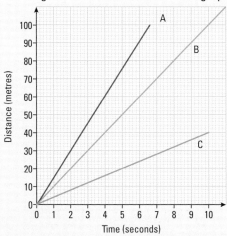

a Work out the gradient of each graph.

b Use your answer to part **a** to match each vehicle to one of the graphs A, B or C.
Give reasons for your answers.

4   Sheila records the depth of water, in cm, in a swimming pool every 30 seconds.
Her results are shown in the table below.

Time (t seconds)	0	30	60	90	120	150	180	210	240
Depth (d cm)	200	175	150	125	100	75	50	25	0

a Draw a straight-line graph to show the depth of water for $t = 0$ to $t = 240$.

b Work out the gradient of the straight line.

c Describe what the gradient represents. What do you think is happening?

9.4 The equation $y = mx + c$

◎ Objectives

○ You can find the gradient of a line from its equation.

○ You can find the y-intercept of a line from its equation.

○ You can find the equation of a straight line.

○ You can interpret the equation of a straight line.

⊘ Why do this?

Straight-line graphs often occur in real life and their equations give the link between two quantities. For example, the equation $F = 1.8C + 32$ can be used to convert °C to °F.

⊕ Get Ready

1. Complete the table to show the gradients of these lines. What do you notice?

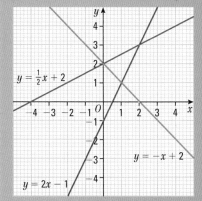

Equation of line	Gradient of line
$y = 2x - 1$	
$y = \frac{1}{2}x + 2$	
$y = -x + 2$	

2. Complete the table to show the y-intercepts of these lines. What do you notice?

Equation of line	y-intercept
$y = 2x - 1$	
$y = \frac{1}{2}x + 2$	
$y = -x + 2$	

Key Points

- The straight line with equation $y = mx + c$ has gradient m.
- The straight line with equation $y = mx + c$ crosses the y-axis at the point $(0, c)$.
- The point $(0, c)$ is known as the y-intercept.

Example 11

For the lines with equations

a $y = 5x + 4$

b $3x + 2y = 6$

find:

    i   the gradient of the line

    ii   the y-intercept of the line.

a $y = 5x + 4$ ← | Compare $y = 5x + 4$ with $y = mx + c$.

i   gradient $= 5$ ← | Write down the value of the gradient m from the term in x.

ii y-intercept $= 4$ ← | Write down the value of the y-intercept c from the constant term.

b i $3x + 2y = 6$ ← | Rearrange the equation $3x + 2y = 6$ into the form $y = mx + c$.

        $2y = 6 - 3x$ ← | Subtract $3x$ from both sides.

        $y = 3 - 1.5x$

        $y = -1.5x + 3$ ← | Divide both sides by 2.

     gradient $= -1.5$

ii y-intercept $= 3$

Example 12

a   Draw the straight line which has a gradient of $\frac{1}{2}$ and which crosses the y-axis at the point $(0, 1)$.

b   Write down the equation of this line.

a

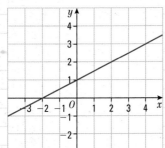

b   Gradient of line $= \frac{1}{2}$

     y-intercept $= 1$

     Equation of the line is $y = \frac{1}{2}x + 1$ ← | Substitute $m = \frac{1}{2}, c = 1$ into $y = mx + c$.

Exercise 9G

1 A line which passes through the point (0, 5) has gradient 2.
Write down the equation of the line.

2 Find **i** the gradient and **ii** the y-intercept of the lines with the equations

   a $y = 4x + 1$ **b** $y = 3x - 4$ **c** $y = \frac{2}{3}x + 4$

   d $2x + 5y = 20$ **e** $4x - 3y = 12$ **f** $x - 2y = 0$

3 Find the equations of the lines shown in the diagram.

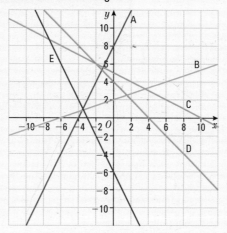

4 A line passes through the points with coordinates (1, 3) and (2, 8). Find the equation of the line.

5 The gradient of a line is 3. The point with coordinates (4, 2) lies on the line.
Find the equation of the line.

9.5 Parallel and perpendicular lines

◉ Objectives

◎ You understand the relationship between the gradients of parallel lines.

◎ You understand the relationship between the gradients of perpendicular lines.

◎ You can use the relationship between the gradients to find the equations of parallel and perpendicular lines.

◈ Why do this?

When building a house you need to find the gradients of parallel and perpendicular lines such as struts and beams in ceilings and roofs.

◈ Get Ready

1. Here are three pairs of lines.
Write down the gradient of each line.

a

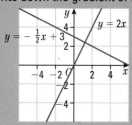

b

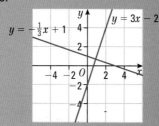

c

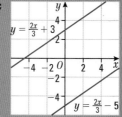

What do you notice?

Key Point

- If a line has gradient m then any line drawn parallel to it also has gradient m and any line drawn **perpendicular** to it has gradient $-\frac{1}{m}$ (the negative reciprocal of m).

Example 13 Find the equation of the line parallel to $y = 3x + 7$ and which passes through the point $(0, -2)$.

$y = 3x + 7$, gradient $m = 3$, y-intercept is $(0, 7)$ ← Compare $y = 3x + 7$ with $y = mx + c$.

The gradient of any line parallel to $y = 3x + 7$ is 3 ← Parallel lines have equal gradients.
so the equation of any line parallel to $y = 3x + 7$ is $y = 3x + c$.

The required line has y-intercept $(0, -2)$. ← Write down the value of c from the coordinates of the y-intercept given.
The equation is $y = 3x - 2$.

Example 14 Find the equation of any line which is perpendicular to $y = 2x - 9$.

$y = 2x - 9$, gradient $m = 2$
The gradient of any line perpendicular to this has gradient $-\frac{1}{2}$. ← Find the negative reciprocal of 2.

The equation of any line with gradient $-\frac{1}{2}$ is of the form $y = -\frac{1}{2}x + c$. ← Use $y = mx + c$.

So $y = -\frac{1}{2}x + 1$ is one example of a line ← Pick any value for c.
perpendicular to the line $y = 2x - 9$.

Example 15 Find the equation of the line which is perpendicular to the line joining the points $(-2, 4)$ and $(4, 1)$ and which passes through the point $(1, 5)$.

Method 1

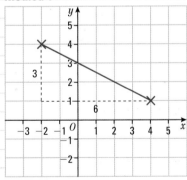

Draw a diagram and find the gradient of the line joining the points $(-2, 4)$ and $(4, 1)$.

Gradient $m = -\frac{3}{6} = -\frac{1}{2}$

The gradient of a perpendicular line is $-\frac{1}{m} = 2$ ← Find the negative reciprocal of $-\frac{1}{2}$.

The equation of any line with gradient 2 is $y = 2x + c$. ← Use $y = mx + c$.

$(1, 5)$ is on this line so
$5 = 2 \times 1 + c$ ← Substitute $x = 1$ and $y = 5$.
$5 = 2 + c$
$c = 3$ ← Find the value of c.

The equation of the line required is $y = 2x + 3$.

Method 2

$$m = \frac{y_2 - y_1}{x_2 - x_1}$$

$$= \frac{1 - 4}{4 - (-2)}$$ ← Use $(-2, 4)$ as (x_1, y_1) and $(4, 1)$ as (x_2, y_2).

$$= \frac{-3}{6}$$

Gradient $= -\frac{1}{2}$

Once the gradient is found, the solution follows **Method 1** above.

Exercise 9H

1. Copy and complete the following table to show the gradients of pairs of lines l_1 and l_2 which are perpendicular to each other.

	a	b	c	d	e
Gradient of line l_1	3	-4	$\frac{1}{5}$		
Gradient of line l_2				3	$-\frac{1}{6}$

2. Write down the equation of a line parallel to the line with the equation
   a $y = 2x + 5$
   b $y = \frac{1}{3}x - 1$
   c $y = 4 - x$

3. Write down the equation of a line perpendicular to the line with the equation
   a $y = x - 6$
   b $y = 3x + 2$
   c $y = 1 - \frac{1}{2}x$

4. Find the equation of a line which is parallel to the line with the equation $y = 4x - 1$ and which passes through the point $(0, 3)$.

5. Find the equation of a line which is parallel to the line with the equation $2x + y = 4$ and which passes through the origin.

6. Find the equation of a line which is perpendicular to the line with the equation $y = \frac{1}{4}x$ and which passes through the point $(2, -8)$.

7. Find the equation of a line which is perpendicular to the line with the equation $x + y = 10$ and which passes through the point $(-2, -5)$.

B

A

9.6 Real-life graphs

- You can draw and interpret distance–time graphs.
- You can draw and interpret other graphs describing real-life situations.

Graphs are used to describe a wide range of real-life situations. You may have to plot and interpret the results of your science class experiment.

◈ **Get Ready**

1. Read off the values shown on the scale.

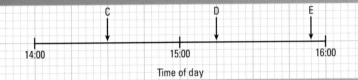

Key Points

- Graphs can be used to describe a variety of real-life situations, and show how one variable changes in relation to another – for example, distance against time, the cost of posting a parcel against its weight, or how a liquid fills a container over time.
- On a **distance–time graph** (or travel graph):
  - straight lines represent **constant speed**
  - horizontal lines represent **no movement**
  - the gradient gives the **speed**: $\text{average speed} = \dfrac{\text{distance travelled}}{\text{time taken}}$

Example 16 ▶ Steve went for a ride on his bike. He rode from his home to a friend's house and back. The travel graph shows his trip.

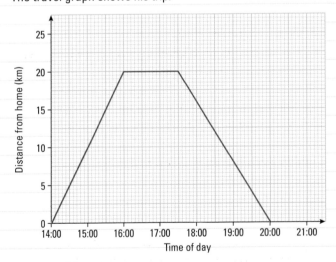

- **a** At what time did Steve reach his friend's house?
- **b** Find Steve's speed on the journey to his friend's house.
- **c** What was Steve doing between 16:00 and 17:30?
- **d** Find Steve's average speed on his journey home.

a Steve reached his friend's house at 16:00 (or 4 pm).

b Speed = $\dfrac{20 \text{ km}}{2 \text{ hours}}$ ← | Work out the gradient of the line representing the first part of his journey. |

| This represents $\dfrac{\text{distance}}{\text{time}}$ = speed |

  = 10 km/h ← | State the units with your answer. |

Results Plus
Examiner's Tip

This may help you to remember the rule to find speed.

Steve travels 10 kilometres each hour on the journey to his friend's house.
Steve's speed is 10 km/h.

c The gradient of the line representing Steve's journey between 16:00 and 17:30 is 0 and so his speed is 0 km/h. He is not moving.
Steve stays at his friend's house between 16:00 and 17:30.

d The gradient of the line representing Steve's journey home is

$\dfrac{20 \text{ km}}{2.5 \text{ hours}}$ ← | The distance home is 20 km. Steve arrives home $2\frac{1}{2}$ hours after he leaves. |

  = 8 km/h

Steve's speed is 8 km/h.

Example 17 ▶ Water is poured into a cylindrical container at a constant rate.

  a Sketch a graph to describe the relationship between the height (h) of the water and the time taken (t).

  b Here is a different-shaped container.
  Sketch a graph showing the relationship between h and t in this case.

a
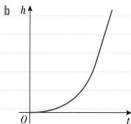

| The container is filled at a constant rate, so the height increases by the same amount for each second. |

| Draw a straight line through $(0, 0)$. |

b

| Since the container gets narrower the height of the water increases more rapidly for each second at first. The gradient of the graph increases. The top part of the container is cylindrical so this is represented by a straight line. |

⚙ **Exercise 9I**

C
A03

1 Janine sets off from home to walk to the shopping
centre. She does some shopping then gets the
bus home. The travel graph shows information
about her journey.

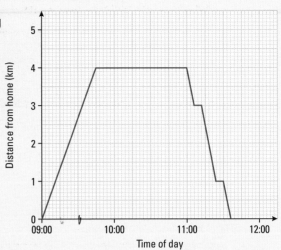

  a What distance from Janine's home is the
    shopping centre?

  b For how many minutes is Janine at the
    shopping centre?

  c Work out Janine's walking speed.
    Give your answer in km/h.

  d The bus stops twice on Janine's journey
    home. For how long does the bus stop
    each time?

  e At what average speed, in km/h, does the bus travel?

A02
A03

2 The graph shows the cost of posting a parcel.

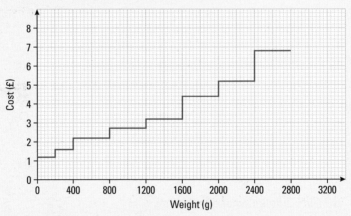

  a Find the cost of posting a parcel of weight 1 kg.

  b Find the maximum weight of a parcel that can be posted for less than £5.

  c Work out the total cost of posting three parcels which have weights 520 g, 1.5 kg and 2.5 kg.

A03

3 Liquid is poured into each of these containers.
Sketch a graph to show the relationship between the depth of water and the volume of water in each
container.

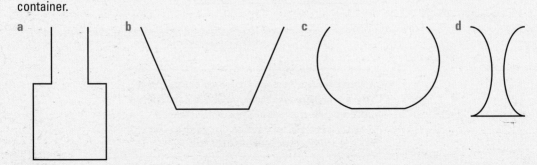

4 Here are the cross-sections of three different swimming pools.
Each pool is to be emptied by pumping out the water. Water is pumped out at a steady rate.

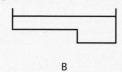

A03 C

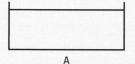

A	B	C

Here are three sketch graphs showing the relationship between the depth of the water left in the pool and the number of minutes since the pump was switched on.

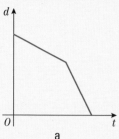
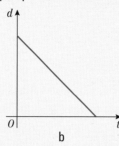
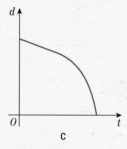

a	b	c

Match each swimming pool with one of the graphs.

5 The petrol consumption of a car, in kilometres per litre (km/l), depends on the speed of the car.
The table gives some information about the petrol consumption of the car at different speeds.

A03

Speed (km/h)	64	72	80	88	96	104
Petrol consumption (km/l)	12.3	13.8	14.4	14.5	14.1	12.3

Draw axes on graph paper, taking 2 cm to represent 10 km/h on the horizontal axis and 4 cm to represent 1 km/l on the vertical axis.
Start the horizontal axis at 60 and the vertical axis at 12.
Plot the values from the table and join them with a smooth curve.
From your graph estimate:
a the petrol consumption at 70 km/h
b the speeds which give a petrol consumption of 14 km/l.

9.7 **Compound measures**

◎ Objective

- You can solve problems with compound measures, giving the correct units.

⊙ Why do this?

Compound measures are used when we want to see how a quantity changes in relation to another quantity, such as by how much the temperature of water increases each second when heated.

⬦ Get Ready

1. How many minutes are there in 1 hour?
2. What fraction of an hour is 15 minutes?
3. What fraction of an hour is 36 minutes? Give your answer as a decimal.
4. How many minutes is 0.3 hours?
5. Write 5.7 hours in hours and minutes.
6. Write 462 minutes in hours and minutes.

Key Points

* A **compound measure** is a measure which involves two units such as km per hour, litres per second or grams per cm³. Compound measures are often a measure of a **rate of change**.

  For example, a tank is filled with 3000 litres of water in 20 minutes. When the rate of filling is constant, this rate means that in 1 minute, $3000 \div 20 = 150$ litres of water would be added to the tank. If the rate is not constant, the average rate of filling the tank is 150 litres in 1 minute.

* In a compound unit, the word 'per' means 'each' or 'for every'. So, the rate of filling the tank is 150 litres per minute or 150 litres/minute. The '/' is like a division sign showing that the rate is the amount of water divided by the time taken.

Example 18   Petrol is leaking from a tank at a rate of 5 litres/hour.

    a Work out how much petrol leaks from the tank in **i** 30 minutes; **ii** 4 hours.

    Initially there are 100 litres of petrol in the tank.

    b Work out how long it takes for all this petrol to leak from the tank.

> **Results**Plus
> **Examiner's Tip**
>
> The units of a compound measure will tell you what to do, so km/l will mean distance ÷ volume.

a   **i** 30 minutes = $\frac{1}{2}$ hour ← *5 litres/hour means in 1 hour 5 litres of petrol leak from the tank.*

      Amount of petrol = $5 \times \frac{1}{2} = 2.5$ litres. ← *Amount = rate × time*

    ii Amount of petrol = $5 \times 4 = 20$ litres.

b Time taken = $100 \div 5 = 20$ hours. ← *Time = amount ÷ rate*

Exercise 9J

1   A car travels 260 km and uses 20 litres of petrol.

   a Work out the average rate of petrol usage. Give your answer in km/litre.

   b Estimate the amount of petrol that would be used when the car has travelled 78 km.

2   A line is turning at a rate of 12° per second.

   a Work out how many degrees the line turns through in 20 seconds.

   b How long does it take for the line to make one complete turn?

3   Water is flowing into a tank. In 5 minutes, 300 litres of water flows in.

   a Work out the average rate of flow of water into the tank. Give the units of your answer.

   b There were 90 litres of water in the tank immediately before the water started to flow in. When full, the tank holds 1200 litres of water. How long will it take for the tank to fill with water? Give your answer in minutes and seconds.

4   On a long journey, a car travels 16 km per litre of petrol. Work out how many litres of petrol the car uses per kilometre.

9.8 Speed

◉ Objective

◉ You can solve problems with average speed.

❓ Why do this?

Speed is a compound measure that we use all the time, from speed limits for cars, to world record breaking running speeds.

◈ Get Ready

1. Work out the average rate of petrol usage if a car travels:
   a 252 km and uses 14 litres of petrol
   b 156 km and uses 24 litres of petrol
   c 84.48 km and uses 6.4 litres of petrol.

🕐 Key Points

◉ Speed is a compound measure because it involves a unit of length and a unit of time, for example kilometres per hour, miles per hour or metres per second. We write kilometres per hour as km/h; the '/' is a sort of division sign showing that speed is distance divided by time.

◉ **Average speed** $= \dfrac{\text{total distance travelled}}{\text{total time taken}}$

If the car travels at an average speed of 30 km/h, the car travels 30 km in 1 hour

$$30 \times 2 = 60 \text{ km in 2 hours}$$
$$30 \times 3 = 90 \text{ km in 3 hours}$$
and so on.

◉ Distance = average speed × time
The time the car takes to travel 90 km at 30 km/h is $\frac{90}{3} = 3$ hours.

Therefore: time $= \dfrac{\text{distance}}{\text{average speed}}$

◉ The following diagram is a useful way to remember these results: D stands for distance, S stands for average speed and T stands for time.

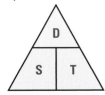

$D = S \times T$

$S = \dfrac{D}{T}$

$T = \dfrac{D}{S}$

🔍 Example 19

The distance from Cardiff to Leeds is 335 km. Rhys drives from Cardiff to Leeds in 6 hours 15 minutes. Work out his average speed for this journey.

Time taken = 6.25 hours ⟵ | Average speed $= \dfrac{\text{total distance travelled}}{\text{total time taken}}$
Time must be in hours.

Average speed $= \frac{335}{6.25}$ ⟵ | 15 minutes $= \frac{15}{60} = 0.25$ hours
The distance is in km and the time is in hours so the speed is in km/h.

$= 53.6$ km/h.

ResultsPlus
Watch Out!

It is important to be careful with time; it is best to use decimals and remember that there are 60 minutes in 1 hour.

Example 20 ▶ Michael decides to go for a cycle ride. He rides a distance of 80 km at an average speed of 24 km/h. Work out how long Michael's ride takes.

Time $= \frac{80}{24} = 3.3333\ldots$ h ⟵

> Time $= \frac{\text{distance}}{\text{average speed}}$
> Speed is in km/h and distance is in km so the time is in hours.

$0.3333\ldots \times 60 = 20$ ⟵

> $3.3333\ldots$ h $= 3$ h $+ 0.3333\ldots$ h
> To change from hours to minutes, multiply by 60.

Time $= 3$ h 20 minutes

Exercise 9K

D

1 Paul takes part in a sponsored hike. He walks 18 km in $4\frac{1}{4}$ hours. What is his average speed? Give your answer correct to 3 significant figures.

C

2 Tim left his home at 11 am and went for a 20 km run. He arrived back at his home at 1 pm. Work out Tim's average speed.

3 A horse runs 12 km at an average speed of 10 km/h. How long, in hours and minutes, does this take?

4 Change a speed of 85 m/s into km/h.

5 In the 2008 Olympics, the men's 100 m race was won in a time of 9.69 s and the men's 200 m race was won in a time of 19.30 s. Which race was won with the faster average speed? You must give a reason for your answer.

Chapter review

- To draw a straight line by plotting points you need to plot at least two points which fit the equation of the line.
- The **equation of a straight line** can have several forms:
  - lines in the form of $x = c$ or $y = c$ where c is a number
  - lines in the form of $y = mx + c$ where m and c are numbers
  - lines in the form of $ax + by = c$ where a, b and c are numbers.
- A line joining two points is called a line segment.
  AB is the line segment joining points A and B.
- The **midpoint** of a line is halfway along the line.
- To find the midpoint you should add the x-coordinates and divide by 2, and add the y-coordinates and divide by 2.
- The midpoint of the line segment AB between A (x_1, y_1) and B (x_2, y_2) is $\left(\dfrac{x_1 + x_2}{2}, \dfrac{y_1 + y_2}{2}\right)$.

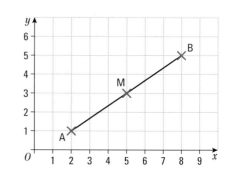

- The **gradient** of a straight line is a measure of its slope.
- Steeper lines have larger gradients.
- Gently sloping lines have smaller gradients.
- Gradient of a line $= \dfrac{\text{change in } y\text{-direction}}{\text{change in } x\text{-direction}}$
- Lines which slope upwards from left to right have positive gradients.
- Lines which slope downwards from left to right have negative gradients.
- The gradient of the line through the points (x_1, y_1) and (x_2, y_2) is given by

$$m = \frac{y_2 - y_1}{x_2 - x_1}$$

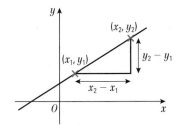

- The **y-intercept** of a line is the value of y when $x = 0$. It is shown by the point where the graph crosses the y-axis.

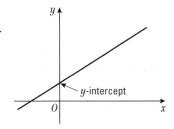

- The straight line with equation $y = mx + c$ has gradient m.
- The straight line with equation $y = mx + c$ crosses the y-axis at the point $(0, c)$.
- The point $(0, c)$ is known as the y-intercept.
- If a line has gradient m then any line drawn parallel to it also has gradient m and any line drawn **perpendicular** to it has gradient $-\frac{1}{m}$ (the negative reciprocal of m).
- On a **distance–time graph** (or travel graph):
  - straight lines represent **constant speed**
  - horizontal lines represent no movement
  - the gradient gives the **speed: average speed** $= \dfrac{\text{distance travelled}}{\text{time taken}}$
- A **compound measure** is a measure which involves two units such as km per hour, litres per second or grams per cm³. Compound measures are often a measure of a **rate of change**.
- In a compound unit, the word 'per' means 'each' or 'for every'. We write litres per minute as litres/minute; the '/' is like a division sign showing that the rate is the amount of liquid divided by the time taken.
- Speed is a compound measure because it involves a unit of length and a unit of time, for example kilometres per hour, miles per hour or metres per second.
- The following diagram is a useful way to remember the relationships between speed, distance and time: D stands for distance, S stands for average speed and T stands for time.

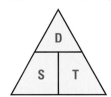

$$\mathbf{D = S \times T}$$
$$\mathbf{S = \frac{D}{T}}$$
$$\mathbf{T = \frac{D}{S}}$$

Review exercise

1   **a**   On the same axes, draw and label the lines $x = 4$, $y = -2$ and $y = -x$.
    b   Work out the area of the triangle formed by the lines $x = 4$, $y = -2$ and $y = -x$.

2   Nicki is going on holiday to the USA.
    She wants to change some pounds (£) to dollars ($). The exchange rate is £1 = $1.65.
    Draw a conversion graph that Nicki could use to change between £ and $.

3   ABCDE is a pentagon.
    a   Find the gradient of:
       i   CD
       ii   BC
       iii   ED.
    b   Use your answers to part **a** to
       write down the gradients of:
       i   AB
       ii   AE.

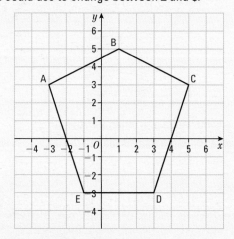

4   The formula $F = 2C + 30$ can be used to estimate F given the value of C, where F is the
    temperature in Fahrenheit and C is the temperature in Celsius.
    Copy and complete the table and use it to draw the graph of F against C for values
    of C from 0 to 100.

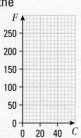

C	0	20	40	60	80	100
F			110			

5   A straight line has equation $2x + y = 6$.
    a   Draw the graph of $2x + y = 6$.
    b   Find:
       i   the gradient
       ii   the y-intercept of the straight line.

6   The distance–time graphs represent the journey made by a bus and a car starting in Swindon,
    travelling to London and returning to Swindon.
    a   How far is it from Swindon to London?
    b   How much longer, including stops, did it
       take the bus to complete the journey
       from Swindon to London than it did the
       car?
    c   Work out the greatest speed of the car
       during the journey.
    d   The bus stopped at Reading on its journey
       to London. At what time did the car reach Reading?
    e   Work out the average speed of the bus, including stops, on its return journey.

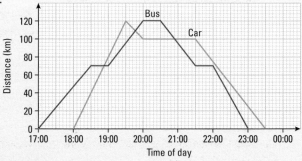

7 Line A has equation $y = 2x + 1$.
Line B passes through the points $(2, -5)$ and $(3, -1)$.
Line C is shown on the diagram.
Which line has the steepest gradient?
Show your working clearly.

A02
A03
C

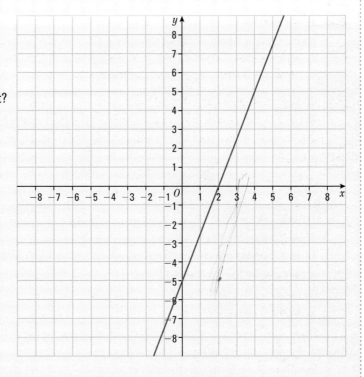

8 Here are four containers.

A   B   C   D

A03

Water is poured into each container at a constant rate.
Sketch a graph for each container showing how the depth of water increases with the volume.

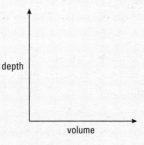

depth

volume

9 The equation of a straight line, l, is $y = \frac{1}{2}x + 3$.

A02

  a Does the point $(2, -1)$ lie on this line?

  b Write down the equation of the line parallel to l which passes through the point $(0, 5)$.

  c Find the equation of the line perpendicular to l which passes through the point $(4, 1)$.

10 Use a suitable grid to draw the graph of $y = 7 - 2x$ for values of x from -1 to 4.

11 The straight line L has the equation $y = 4x + 3$.

  a State the gradient of this line.

  b Find the coordinates of the point where L cuts the x-axis.

  The point $(k, 21)$ lies on L.

  c Find the value of k.

12 The graph shows the cost of using a mobile phone for one month for three different tariffs.

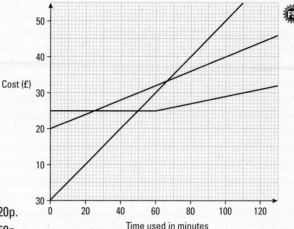

Cost (£)

Time used in minutes

Tariff A   Rental £20   Every minute costs 20p.
Tariff B   Pay as you go   Every minute costs 50p.
Tariff C   Rental £25   First 60 minutes free, then each minute costs 10p.

a Label each line on the graph with the letter of the tariff it represents.

Jim uses tariff A for 100 minutes in one month.

b Find the total cost.

Fiona uses her mobile phone for about 60 minutes each month.

* c Explain which tariff would be the cheapest for her to use.

You must give the reasons for your answer.

* **13** Abbie has the option of joining two health clubs.

Hermes has a joining fee of £100 plus a fee of £5 per session.

Atlantis has a joining fee of £200 with a fee of £3 per session.

Which health club should she choose?

You must show all calculations and fully explain your solution.

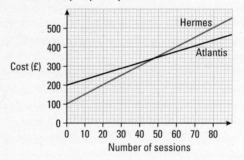

Cost (£)

Number of sessions

14 P has coordinates (1, 4).

R has coordinates (5, 0).

Find the coordinates of the midpoint of the line PR.

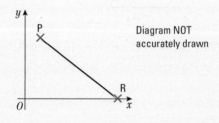

Diagram NOT accurately drawn

June 2008

15 Stuart drives 180 km in 2 hours 15 minutes.

Work out Stuart's average speed.

Nov 2008

16 The distance from London to New York is 3456 miles.

A plane takes 8 hours to fly from London to New York.

Work out the average speed of the plane.

June 2008

17 John travelled 30 km in 1.5 hours.

Kamala travelled 42 km in 2 hours.

Who had the greater average speed?

You must show your working.

June 2009

18 There are 40 litres of water in a barrel.

The water flows out of the barrel at a rate of 125 millilitres per second.

1 litre = 1000 millilitres

Work out the time it takes for the barrel to empty completely.

June 2009

19 Copy and complete the following table.

Equation of line	Gradient	y-intercept
$y = 2x + 5$		
	7	−3
$y = 6 - x$		
	$\frac{2}{3}$	−1
	−4	3

20 Here are six temperature/time graphs.

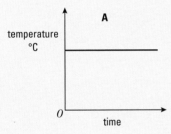

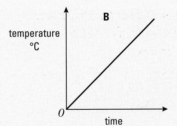

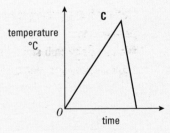

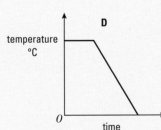

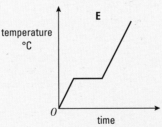

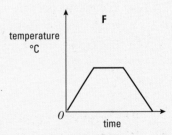

Describe the events shown by the graph in each case.

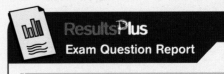

ResultsPlus

Exam Question Report

98% of students answered this sort of question well.

Nov 2008, adapted

A

A03

21 The diagram shows a rectangle.
All the measurements are in centimetres.
The perimeter of the rectangle is 24 cm.

2x

y

Diagram NOT
accurately drawn

a Explain why $2x + y = 12$.
b Draw the graph of $2x + y = 12$ for values of x from 0 to 6.
c Use your graph to find the value of x which makes the rectangle a square.

A02

22 The point P $(3, k)$ lies on the line with equation $y = 2x + 1$.
Show that P also lies on the line with equation $y = 3x - 2$.

A03

23 The diagram shows three points $A (-1, 5)$, $B (2, -1)$ and $C (0, 5)$.
The line L is parallel to AB and passes through C.
Find the equation of the line L.

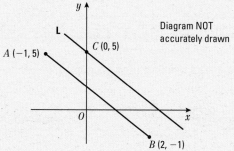

June 2005

A02

24 A straight line has equation $y = 2x - 3$.
The point P lies on the straight line.
The y coordinate of P is -4.
a Find the x coordinate of P.
A straight line L is parallel to $y = 2x - 3$ and passes through the point $(3, 4)$.
b Find the equation of the line L.

Nov 2005

One of the most famous formulae you will come across is Einstein's $e = mc^2$ from his theory of relativity. Einstein's brain was removed after his death and researchers in Canada compared it with the brains of 91 people of average intelligence to try to discover the secret of his outstanding intelligence. They found that the area of Einstein's brain that is responsible for mathematical thought and spatial awareness was much larger, and his brain was 15% wider than the others.

◉ Objectives

In this chapter you will:
- ◉ distinguish between the words 'equation', 'formula', 'identity' and 'expression'
- ◉ use and derive algebraic formulae.

◈ Before you start

You should already know how to:
- ◉ collect like terms in an algebraic expression
- ◉ know how to substitute into algebraic expressions.

10.1 Distinguishing between 'equation', 'formula', 'identity' and 'expression'

◎ Objective

○ You can distinguish between the words 'equation', 'formula', 'identity' and 'expression'.

◈ Why do this?

You need to understand mathematical words to be able to understand exam questions.

⬆ Get Ready

1. Evaluate these expressions when $x = 4$ and $y = -2$.

a $x - y$ b $x + y$ c $4 + x + 2y$

🔍 Key Points

⦿ An example of an algebraic expression is $2p - 10$. It is made up of the terms $2p$ and -10.
⦿ An example of an **equation** is $2p - 10 = 9$. This can be solved to find the value of p.
⦿ $2(p - 5) = 2p - 10$ is called an **identity**. The left-hand side is the same as the right-hand side.
⦿ $C = 2p - 10$ is called a **formula**. If the value of p is known, you can substitute it into the equation to work out the value of C.

🔍 Example 1

Write down whether each of the following is an expression, an equation, an identity or a formula.

a $5ab - 2ab = 3ab$

b $5p + 2q$

c $T = 2\pi\sqrt{\dfrac{l}{g}}$

d $\dfrac{1 - 2x}{3} = 5$

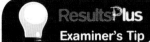

Results Plus
Examiner's Tip

An expression is the only one without an '=' sign.

a The right-hand side is the same as the left-hand side and so $5ab - 2ab = 3ab$ is an identity.   ←   Collect like terms of $5ab - 2ab$ to give $3ab$.

b $5p + 2q$ is an expression.

c If the values of l and g are known, the value of T can be worked out using the formula $T = 2\pi\sqrt{\frac{l}{g}}$.

d $\dfrac{1 - 2x}{3} = 5$, can be solved to find the value of x and is therefore an equation.

⚙ Exercise 10A

Write down whether each of the following is an expression, an equation, an identity or a formula.

1   $A = 2(l + b)$ 2 $m + m + m + m = 4m$ 3 $2a + 3b$

4   $3y^2 = 243$ 5 $y = mx + c$ 6 $E = mc^2$

7	$x^2 - 5x$	8	$\dfrac{1}{x} + \dfrac{2}{3x} = 4$	9	$V = \dfrac{4}{3}\pi r^3$
10	$2(x + 3) = x + 4$	11	$x^2 - 5x = x(x - 5)$		
12	$x^2 \times x^5 = x^7$				

D

C

10.2 Using formulae

◎ Objective

○ You can use formulae from mathematics and other subjects.

◈ Why do this?

Mobile phone bills are calculated using a formula which is based on the number and length of calls made and number of texts sent.

⬆ Get Ready

1. Work out   **a** 64.9×82.4 **b** $\sqrt{7\tfrac{1}{9}}$

🔍 Key Points

◉ A formula is a way of describing a relationship between two or more sets of values.
Area = length × width is an example of a **word formula**.

◉ A formula is written using words or algebraic expressions.
Einstein's theory of relativity is described by the **algebraic formula** $\boxed{E = mc^2}$

🔍 Example 2

The area of a rectangle is given by the formula.

Area of a rectangle = length × width.

    a Find the area of a rectangle of length 8 cm and width 5 cm.

    b Find the length of a rectangle with an area of 27 mm² and width of 10 mm.

a Length = 8 cm, width = 5 cm   ← | Substitute the values for the length and the width.
Area = length × width = 8 cm × 5 cm
Area = 40 cm²

ResultsPlus

Examiner's Tip

The first step to finding a value from a formula is to simply replace each word by its value.

b Area = length × width   ← | Divide both sides of the equation by 10.
27 = length × 10
Length = 27 ÷ 10
       = 2.7 mm

🔍 Example 3

Use the formula $E = mc^2$ to work out the value of E when $m = 4$ and $c = 3$.

$E = mc^2$ ← | Substitute $m = 4$ and $c = 3$ into the formula.
$E = 4 \times 3^2$
   $= 4 \times 9$
$E = 36$

Exercise 10B

1 $v = u + at$

Work out the value of v when

a $u = 80$, $a = 10$ and $t = 4$ **b** $u = 35$, $a = -5$ and $t = 12$

2 $T = 3p^2 - 2p$

Work out the value of T when   **a** $p = 5$ **b** $p = -1$

3 $V = \frac{1}{3}\pi r^2 h$

Work out the value of V when

a $\pi = 3.14$, $r = 10$ and $h = 15$ **b** $\pi = 3.14$, $r = 2.4$ and $h = 20$

4 Use the formula **distance** = **speed** $\times$ **time** to work out:

a the distance travelled by a car travelling for $2\frac{1}{2}$ hours at an average speed of 48 mph.

b the average speed of an athlete running 100 metres in 12.5 seconds.

5 The cooking time, in minutes, of a turkey is given by the following formula:

cooking time = **weight of turkey in lb** $\times$ **25** + **30**.

Use this formula to work out:

a the cooking time for a turkey weighing 11 lb, giving your answer in hours and minutes

b the weight, in lb, of a turkey taking 8 hours to cook.

6 $a = \sqrt{c^2 - b^2}$

Work out the value of a when

a $c = 13$ and $b = 5$ **b** $c = 41$ and $b = 40$

10.3 Deriving an algebraic formula

◎ Objective

○ You can derive an algebraic formulae from information given.

◈ Why do this?

Chemists need to derive algebraic formulae when producing the correct balance of ingredients in medicines.

◈ Get Ready

1. $V = I^2 R$

Work out V when

a $I = 4$, $R = 200$ **b** $I = 0.3$, $R = 100$

◉ Key Points

◉ You can use information given to form an algebraic expression. You give each variable a letter.

◉ You must define each variable.

Example 4

A farmer sells sheep and cows at the local market. Each sheep is sold for £s and each cow is sold for £c.

a Write down a formula for his total sales £T, in terms of s and c, if he sells 120 sheep and 45 cows.

b Find the value of T if $s = 150$ and $c = 480$.

ResultsPlus

Watch Out!

Do not try to combine the terms if the variables are different.

a Income from sale of 120 sheep is $120 \times s = 120s$.
Income from sale of 45 cows is $45 \times c = 45c$.
Total sales is $120s + 45c$.
The formula is therefore $T = 120s + 45c$.

← The total is required, so add the expressions.

b $s = 150$ and $c = 480$

← Substitute given values into the formula.

Substituting into $T = 120s + 45c$

$T = 120 \times 150 + 45 \times 480$

$T = 18\,000 + 21\,600$

$T = 39\,600$

Exercise 10C

1. Duncan hires a car whilst on holiday in Spain. The cost of hiring a car is €90 plus €50 for each day that the car is hired for.
   a Write down a formula that could be used to find the total cost, €C, to hire a car for d days.
   b Use your formula to work out the cost of hiring a car for 14 days.

   D

2. In some games, 5 points are awarded for a win, 3 points are awarded for a draw and 1 point is awarded for a loss.
   In one evening, Caroline wins x games, draws y games and loses z games.
   Write down a formula, in terms of x, y and z, for the total points (P) scored by Caroline.

3. David owns a hairdressing salon. The average length of time spent on a male client is 45 minutes.
   The average length of time spent on a female client is 75 minutes.
   In one week David had m male clients and f female clients.
   Write down a formula, in terms of m and f, for the total time, T hours, that David spent on his clients during this week.

   C
   A02

4. The diagram shows the plan of an L-shaped room.
   The dimensions of the room are given in metres.
   Write down a formula, in terms of x and y,
   for the perimeter, P metres, of the room.

   A02

   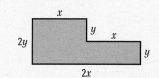

5. Adult cinema tickets cost £a and child cinema tickets cost £c.
   Mr Brown buys 2 adult tickets and 5 child tickets.
   a Write down a formula, in terms of a and c, for the total cost, £T, of these tickets.

   The following week the cinema has a special offer.
   'For each adult ticket bought, one child ticket is free.'
   Mr Brown again buys 2 adult tickets and takes the same 5 children.
   b Write down a formula, in terms of a and c, for the new total cost, £P, of these tickets.

   A02

Chapter review

- An algebraic expression is made up of terms.
- An **equation** can be solved to find the value of a term.
- In an **identity** the left-hand side is the same as the right-hand side.
- If the value of a term is known, you can substitute it into a **formula** to work out the value of another term.
- A formula is a way of describing a relationship between two or more sets of values.
- A **word formula** is written in words.
- An **algebraic formula** is written using algebraic expressions.
- You can use information given to form an algebraic expression. You give each variable a letter.
- You must define each variable.

Review exercise

1. Write down whether each of the following is an expression, an equation, an identity or a formula.

   a $\dfrac{3p}{q} = 3pq^{-1}$ b $\frac{1}{2}a^2 - b$ c $F = mg$

   d $m^n + n^m$ e $12 - s^3 = 36$ f $\sqrt[3]{t} = t^{\frac{1}{3}}$

2. $P = 4c + d$

   Work out the value of P when

   a $c = 11$ and $d = 7$

   b $c = -3$ and $d = 9$

3. Compasses cost c pence each. Rulers cost r pence each.

   Write down an expression for the total cost, in pence, of 2 compasses and 4 rulers.   *June 2009*

4. Colin is three years younger than Ben.

   a Write down an expression for Colin's age.

   Daniel is twice as old as Ben.

   b Write down an expression for Daniel's age.   *Nov 2008*

5. $y = 7 - 4x$

   Work out the value of y when   a $x = 5$ b $x = -2$

6. $f = \dfrac{uv}{u + v}$

   Work out the value of f when   a $u = 3$ and $v = 4$,   b $u = -10$ and $v = 20$

7. $v^2 = u^2 + 2as$ $u = 6, a = 3, s = 7.5$

   Work out the value of v.   *Nov 2008*

8. Adam, Barry and Charlie each buy some stamps.

   Barry buys three times as many stamps as Adam.

   a Write down an expression for the number of stamps Barry buys.

   Charlie buys 5 more stamps than Adam.

   b Write down an expression for the number of stamps Charlie buys.

9 The cost of hiring a car can be worked out using this rule.

Cost = £90 + 50p per mile

Bill hires a car and drives 80 miles.

a Work out the cost.

The cost of hiring a car and driving m miles is C pounds.

b Write the formula for C in terms of m.

Nov 2007

10 $y = \dfrac{a^2 - c^2}{a^2 + c^2}$ $a = 3.2,$ $c = 1.6$

Work out the value of y.

11 $s = ut + \frac{1}{2}at^2$

Work out the value of s when

a $u = 30$, $t = 3.5$ and $a = 10$

b $u = 5$, $t = 4$ and $a = -2$

11 ALGEBRAIC FRACTIONS AND ALGEBRAIC PROOF

The word algebra is derived from the Arabic word Al-Jabr. It first appeared in 820AD in the work of the Persian mathematician Al-Khwarizmi. Al-Khwarizmi became known as the father of algebra, creating and using algebraic proofs to solve the mathematical problems of the time.

Objectives

In this chapter you will:
- simplify algebraic fractions
- add and subtract algebraic fractions
- multiply and divide algebraic fractions
- prove a given result using algebra.

Before you start

You need to be able to:
- add, subtract, multiply and divide fractions
- factorise algebraic expressions
- use the laws of indices.

11.1 Simplifying algebraic fractions

Objective

○ You can simplify algebraic fractions.

Why do this?

Doctors, engineers and scientists often have to use and simplify algebraic fractions in their jobs.

Get Ready

1. Factorise fully.

   a $x^2 + 5x + 4$ b $x^2 - x - 6$ c $2x^2 + 7x + 3$

Key Points

● **Algebraic fractions**, like numerical fractions, can often be simplified.

● To simplify an algebraic fraction, factorise the numerator and the denominator. Then divide the numerator and denominator by any common factors.

Example 1

Simplify fully $\dfrac{2x^2 + 4x}{x^2 + 3x + 2}$

$2x^2 + 4x = 2x(x + 2)$ ← Factorise the numerator fully.

$x^2 + 3x + 2 = (x + 2)(x + 1)$ ← Factorise the denominator.

$\dfrac{2x^2 + 4x}{x^2 + 3x + 2} = \dfrac{2x(x + 2)^1}{{}^1(x + 2)(x + 1)}$ ← Write the fraction in fully factorised form.

$= \dfrac{2x}{(x + 1)}$ ← Divide both the numerator and denominator by the common factor $(x + 2)$.

$= \dfrac{2x}{x + 1}$ ← Write your answer without the bracket as there is only one factor in the denominator.

ResultsPlus

Watch Out!

You cannot simplify an algebraic fraction until it has been expressed as a product of factors.

Example 2

Simplify fully $\dfrac{2x^2 - 5x - 3}{x^3 - 9x}$

$2x^2 - 5x - 3 = (2x + 1)(x - 3)$ ← Factorise the numerator.

$x^3 - 9x = x(x^2 - 9)$ ← Factorise the denominator.

$= x(x + 3)(x - 3)$ ← Take out the common factor x. Factorise $(x^2 - 9)$ using the difference of two squares.

$\dfrac{2x^2 - 5x - 3}{x^3 - 9x} = \dfrac{(2x + 1)(x - 3)}{x(x + 3)(x - 3)}$ ← Write the fraction in factorised form.

$= \dfrac{(2x + 1)(x - 3)^1}{x(x + 3)(x - 3)^1}$ ← Divide both the numerator and denominator by any common factors.

$= \dfrac{2x + 1}{x(x + 3)}$ ← Write the numerator without the bracket as there is only one factor left in the numerator.

Exercise 11A

Questions in this chapter are targeted at the grades indicated.

1 Simplify fully.

a $\dfrac{2x^5}{x^2}$ b $\dfrac{x^2y}{3xy^2}$ c $\dfrac{x^2 - 5x}{2x}$ d $\dfrac{x^2 + 3x}{x + 3}$ e $\dfrac{2x - 4x^2}{2x - 1}$

2 Simplify fully.

a $\dfrac{x^2 + 4x + 3}{x^2 + 5x + 6}$ b $\dfrac{x^2 + 6x + 5}{x^2 + 5x}$ c $\dfrac{x^2 - 5x + 6}{x^2 + x - 12}$ d $\dfrac{x^2 - x - 12}{x^2 + 6x + 9}$

3 Simplify fully.

a $\dfrac{x^2 - 1}{x^2 - x}$ b $\dfrac{4x^2 + 24x}{x^2 - 36}$ c $\dfrac{2x^2 - 8}{x^2 + 4x + 4}$ d $\dfrac{3x^2 - 27}{3x^2 + 9x}$

4 Simplify fully.

a $\dfrac{2x^2 + 5x + 3}{3x^2 + 5x + 2}$ b $\dfrac{10x^2 - x - 3}{6x^2 - x - 2}$ c $\dfrac{9x^2 - 1}{9x^2 - 6x + 1}$ d $\dfrac{6x^2 + 5x - 1}{12x^2 + 16x - 3}$

5 Simplify fully.

a $\dfrac{x(x + 5)}{x^2 - 5x}$ b $\dfrac{x^2 - 10x + 25}{2x^2 - 50}$ c $\dfrac{8x^2 - 10x + 3}{8x^2 - 6x}$

d $\dfrac{6x^2 - 2x^3}{2x^3 + 6x^2}$ e $\dfrac{4 - x^2}{(x + 2)^2}$ f $\dfrac{16 - x^2}{x - 4}$

11.2 Adding and subtracting algebraic fractions

◎ Objective

● You can add and subtract algebraic fractions.

◈ Why do this?

If you win two competitions in a tennis tournament, and you know what fraction of the prize money each one is worth, you can work out what the total prize money for the tournament was.

◈ Get Ready

1. Write down the lowest common multiple (LCM) of:
   a 6 and 15   b $3x$ and $4x$ c $(x + 1)$ and $x(x + 1)$.

◔ Key Points

● To add (or subtract) algebraic fractions we use a similar method to that used for adding and subtracting numerical fractions.

● If the denominators of the fractions are the same, add (or subtract) the numerators but do not change the denominator.

● To add (or subtract) algebraic fractions with different denominators, find a common denominator and write each fraction as an equivalent fraction with this denominator.

● To find the lowest common denominator of algebraic fractions, you may need to factorise the denominators first.

● To simplify your answers, you may have to factorise the numerator.

Example 3 Add $\dfrac{2}{x} + \dfrac{3}{x}$

$\dfrac{2}{x} + \dfrac{3}{x} = \dfrac{5}{x}$ ← The denominators are the same so just add the numerators.

Example 4 Subtract $\dfrac{5x}{7} - \dfrac{3x}{7}$

$\dfrac{5x}{7} - \dfrac{3x}{7} = \dfrac{2x}{7}$ ← Subtract the numerators.

Example 5 Write $\dfrac{3}{2x} - \dfrac{1}{x}$ as a single fraction.

$\dfrac{3}{2x} - \dfrac{1}{x} = \dfrac{3}{2x} - \dfrac{2}{2x}$ ← Write each fraction with the same common denominator.

$= \dfrac{1}{2x}$ ← Subtract the numerators, but leave the denominator the same.

Example 6 Simplify $\dfrac{x+2}{3} + \dfrac{x-1}{4}$

Common denominator = 12. ← Work out the lowest common denominator.

$\dfrac{x+2}{3} + \dfrac{x-1}{4} = \dfrac{4(x+2)}{12} + \dfrac{3(x-1)}{12}$ ← Write as equivalent fractions with the same denominator.

$= \dfrac{4(x+2) + 3(x-1)}{12}$ ← Add the two fractions.

$= \dfrac{4x + 8 + 3x - 3}{12}$ ← Expand the brackets.

$= \dfrac{7x + 5}{12}$ ← Simplify the numerator.

Example 7 Write $\dfrac{3}{x-1} - \dfrac{2}{x+1}$ as a single fraction.

Common denominator = $(x-1)(x+1)$ ← Find a common denominator.

$\dfrac{3}{x-1} - \dfrac{2}{x+1} = \dfrac{3(x+1)}{(x-1)(x+1)} - \dfrac{2(x-1)}{(x-1)(x+1)}$ ← Convert each fraction to an equivalent fraction with the common denominator $(x-1)(x+1)$.

$= \dfrac{3(x+1) - 2(x-1)}{(x-1)(x+1)}$ ← Subtract the fractions.

$= \dfrac{3x + 3 - 2x + 2}{(x-1)(x+1)}$ ← Expand the brackets in the numerator.

$= \dfrac{x + 5}{(x-1)(x+1)}$ ← Simplify the numerator.

ResultsPlus
Watch Out!

There is no need to multiply out the brackets in the denominator.

Exercise 11B

1 Write as a single fraction in its simplest form.

a $\dfrac{2x}{3} + \dfrac{x}{3}$ b $\dfrac{x}{2} + \dfrac{3x}{2}$ c $\dfrac{3}{10x} + \dfrac{4}{10x}$

d $\dfrac{7x}{9} - \dfrac{3x}{9}$ e $\dfrac{4x}{5} - \dfrac{3x}{5}$ f $\dfrac{7}{3x} - \dfrac{2}{3x}$

2 Write as a single fraction in its simplest form.

a $\dfrac{x}{3} + \dfrac{x}{4}$ b $\dfrac{x}{5} + \dfrac{2x}{15}$ c $\dfrac{x}{2} - \dfrac{x}{8}$

d $\dfrac{3x}{2} - \dfrac{2x}{3}$ e $\dfrac{1}{3x} + \dfrac{1}{2x}$ f $\dfrac{4}{10x} - \dfrac{3}{20x}$

3 Simplify.

a $\dfrac{x}{2} + \dfrac{x+1}{3}$ b $\dfrac{x-3}{4} + \dfrac{x+2}{5}$ c $\dfrac{2x}{3} - \dfrac{5x}{9}$

d $\dfrac{1}{x+2} + \dfrac{1}{x+3}$ e $\dfrac{4}{x+2} - \dfrac{3}{x+1}$ f $\dfrac{1}{2x-1} - \dfrac{1}{2x+3}$

Example 8   Write $\dfrac{1}{x} - \dfrac{3}{x^2 + 3x}$ as a single fraction in its simplest form.

$\dfrac{1}{x} - \dfrac{3}{x^2 + 3x} = \dfrac{1}{x} - \dfrac{3}{x(x+3)}$ ← Factorise the denominator $x^2 + 3x$.

Common denominator $= x(x+3)$ ← Find the lowest common denominator.

$\dfrac{1}{x} - \dfrac{3}{x^2 + 3x} = \dfrac{x+3}{x(x+3)} - \dfrac{3}{x(x+3)}$ ← Write each fraction with the same denominator.

$= \dfrac{x+3-3}{x(x+3)}$ ← Combine the fractions.

$= \dfrac{x}{x(x+3)}$ ← Simplify the numerator.

$= \dfrac{1}{x+3}$ ← Divide the numerator and denominator by x to simplify your answer.

ResultsPlus Examiner's Tip

$x \div x = 1$
$x(x+3) \div x = x+3$

Example 9   Simplify $\dfrac{1}{5x+10} + \dfrac{1}{x^2 + 5x + 6}$

$5x + 10 = 5(x+2)$
$x^2 + 5x + 6 = (x+3)(x+2)$ ← Factorise each denominator.

$\dfrac{1}{5x+10} + \dfrac{1}{x^2+5x+6} = \dfrac{1}{5(x+2)} + \dfrac{1}{(x+3)(x+2)}$ ← Replace the denominators with the factorised expressions.

Common denominator $= 5(x+3)(x+2)$ ← Find the lowest common denominator of $5(x+2)$ and $(x+3)(x+2)$.

$\dfrac{1}{5x+10} + \dfrac{1}{x^2+5x+6} = \dfrac{(x+3)}{5(x+3)(x+2)} + \dfrac{5}{5(x+3)(x+2)}$ ← Write equivalent fractions with the common denominator.

$= \dfrac{x+3+5}{5(x+3)(x+2)}$ ← Combine the fractions.

$= \dfrac{x+8}{5(x+3)(x+2)}$

Exercise 11C

1 a Factorise   i $2x + 2$ ii $6x + 6$.
   b Write down the lowest common multiple of $2x + 2$ and $6x + 6$.
   c Write $\dfrac{1}{2x + 2} + \dfrac{1}{6x + 6}$ as a single fraction in its simplest form.

2 Write down the lowest common multiple of each of the following pairs of expressions.
   a $3x$ and $5x$ b $x + 2$ and $x + 3$
   c x and $x(x - 1)$ d $x + 2$ and $(x + 1)(x + 2)$
   e $2x - 6$ and $x - 3$ f $x + 1$ and $x^2 + x$

3 a Factorise $x^2 + 3x + 2$.
   b Write $\dfrac{1}{x + 2} - \dfrac{1}{x^2 + 3x + 2}$ as a single fraction in its simplest form.

4 a Factorise $x^2 - 4$.
   b Write $\dfrac{3}{x - 2} - \dfrac{2}{x^2 - 4}$ as a single fraction in its simplest form.

5 a Factorise $2x^2 - 3x + 1$.
   b Write $\dfrac{1}{2x^2 - 3x + 1} + \dfrac{2}{2x - 1}$ as a single fraction in its simplest form.

6 Simplify $\dfrac{1}{2x + 6} - \dfrac{1}{x^2 + 4x + 3}$

7 Write $\dfrac{1}{4} + \dfrac{1}{2x} + \dfrac{1}{8(x + 1)}$ as a single fraction.

8 Express $\dfrac{3}{3 - x} - \dfrac{9}{9 - x^2}$ as a single fraction.

9 a Factorise   i $x^2 + 9x + 20$ ii $x^2 + 11x + 30$.
   b Write $\dfrac{4}{x^2 + 9x + 20} - \dfrac{1}{x^2 + 11x + 30}$ as a single fraction in its simplest form.

10 Show that $\dfrac{1}{4x^2 - 8x + 3} - \dfrac{1}{4x^2 - 1} = \dfrac{A}{(2x - 1)(2x + 1)(2x - 3)}$ and find the value of A.

D

B

A*

11.3 Multiplying and dividing algebraic fractions

◎ Objective

• You can multiply and divide algebraic fractions.

◈ Why do this?

Doctors and nurses need to multiply and divide algebraic fractions when calculating the drugs dosage to give their patients.

◈ Get Ready

1. Simplify.

   a $\dfrac{ab}{b}$ b $\dfrac{x + 1}{2x + 2}$ c $\dfrac{(x + 1)(x + 3)}{(x + 1)^2}$

🔍 Key Points

- To multiply (or divide) algebraic fractions we use a similar method to that used for multiplying and dividing numerical fractions.
- To multiply fractions, multiply the numerators and multiply the denominators.
$$\frac{a}{b} \times \frac{c}{d} = \frac{ac}{bd}$$
- To divide fractions, multiply the first fraction by the reciprocal of the second.
$$\frac{a}{b} \div \frac{c}{d} = \frac{a}{b} \times \frac{d}{c} = \frac{ad}{bc}$$
- Simplify your answers if you can.
- You may need to factorise the numerator and/or the denominator before you multiply or divide algebraic fractions.

Example 10   Simplify $\dfrac{2x}{3} \times \dfrac{x}{4}$

$$\frac{2x}{3} \times \frac{x}{4} = \frac{2x \times x}{3 \times 4} = \frac{{}^{1}\cancel{2}x^2}{\cancel{12}_{6}}$$

← Multiply $2x$ by x. Work out 3×4.

$$= \frac{x^2}{6}$$

← Divide both the numerator and denominator by 2.

Example 11   Simplify $\dfrac{2x}{5y} \div \dfrac{x^2}{y}$

$$\frac{2x}{5y} \div \frac{x^2}{y} = \frac{2x}{5y} \times \frac{y}{x^2}$$

← Multiply the first fraction by the reciprocal of the second.

$$= \frac{2 \times \overset{1}{\cancel{x}} \times \overset{1}{\cancel{y}}}{5 \times \underset{1}{\cancel{y}} \times \underset{1}{\cancel{x}} \times x}$$

← Divide both the numerator and denominator by x and by y.

$$= \frac{2}{5x}$$

Example 12   Simplify $\dfrac{x+1}{x+2} \times \dfrac{(x+2)^2}{3(x+1)}$

$$\frac{x+1}{x+2} \times \frac{(x+2)^2}{3(x+1)} = \frac{\overset{1}{\cancel{(x+1)}}\overset{1}{\cancel{(x+2)}}(x+2)}{3\underset{1}{\cancel{(x+2)}}\underset{1}{\cancel{(x+1)}}}$$

← Write down the product of the two numerators and the product of the two denominators.

$$= \frac{x+2}{3}$$

← Simplify the fraction.

Example 13   Simplify $\dfrac{2x-1}{4} \div \dfrac{4x-2}{5}$

$$\frac{2x-1}{4} \div \frac{4x-2}{5} = \frac{2x-1}{4} \div \frac{2(2x-1)}{5}$$

← Factorise the numerator of the second fraction.

$$= \frac{2x-1}{4} \times \frac{5}{2(2x-1)}$$

$$= \frac{5\overset{1}{\cancel{(2x-1)}}}{8\underset{1}{\cancel{(2x-1)}}}$$

← Divide the numerator and denominator by $2x-1$.

$$= \frac{5}{8}$$

Example 14 Simplify $\dfrac{2x + 1}{x^2 - 1} \times \dfrac{x + 1}{2x^2 - x - 1}$

$\dfrac{2x + 1}{x^2 - 1} \times \dfrac{x + 1}{2x^2 - x - 1} = \dfrac{2x + 1}{(x - 1)(x + 1)} \times \dfrac{x + 1}{(2x + 1)(x - 1)}$

← Factorise $x^2 - 1$ and $2x^2 - x - 1$.

$= \dfrac{{}^{1}(2x + 1)(x + 1)^{1}}{(x - 1)(x + 1)(2x + 1)(x - 1)}$

← Write down the product of the two fractions and simplify.

$= \dfrac{1}{(x - 1)(x - 1)}$

$= \dfrac{1}{(x - 1)^2}$

Write $(x - 1)(x - 1)$ as $(x - 1)^2$.

ResultsPlus
Examiner's Tip

Note that it is often preferable to leave the fraction in factorised form.

Exercise 11D

1 Write as a single fraction.

a $\dfrac{x}{3} \times \dfrac{x}{5}$ b $\dfrac{4}{y} \times \dfrac{3}{y}$ c $\dfrac{5x}{2} \times \dfrac{3y}{4}$ d $\dfrac{x}{3} \times \dfrac{x - 3}{4}$

2 Write as a single fraction in its simplest form.

a $\dfrac{3x}{5} \times \dfrac{10y}{12}$ b $\dfrac{4x}{9y} \times \dfrac{y}{2}$ c $\dfrac{2x^2}{y^2} \times \dfrac{3y}{x^2}$ d $\dfrac{x + 1}{x} \times \dfrac{2x}{x - 1}$

3 Write as a single fraction.

a $\dfrac{x}{9} \div \dfrac{x}{5}$ b $\dfrac{5x}{6} \div \dfrac{3}{y}$ c $x^2y \div \dfrac{1}{y}$ d $\dfrac{2x}{x + 1} \div \dfrac{x + 1}{x + 2}$

4 Write as a single fraction in its simplest form.

a $\dfrac{3x}{2} \div \dfrac{2x}{9}$ b $\dfrac{5y^2}{8x} \div \dfrac{y^2}{x^2}$ c $\dfrac{7x}{12y} \div \dfrac{y}{6}$ d $\dfrac{x}{2} \div \dfrac{x - 5}{4}$

5 Write as a single fraction in its simplest form.

a $\dfrac{x + 1}{3} \times \dfrac{3x + 3}{2}$ b $\dfrac{x + 2}{x - 1} \times (x - 1)^2$ c $\dfrac{x}{x + 1} \times \dfrac{x + 1}{x + 2}$

d $\dfrac{x + 4}{9} \div \dfrac{2x + 8}{3}$ e $\dfrac{6}{3x - 1} \div \dfrac{2}{(3x - 1)^2}$ f $\dfrac{3x - 12}{4} \div \dfrac{x - 4}{x + 4}$

6 a Factorise $x^2 - 4$.

b Write $\dfrac{1}{x + 2} \times \dfrac{x^2 - 4}{x^2 + 4}$ as a single fraction in its simplest form.

7 a Factorise   i $x^2 + 5x + 4$ ii $x^2 + 6x + 8$.

b Write $\dfrac{x + 3}{x^2 + 5x + 4} \div \dfrac{x + 1}{x^2 + 6x + 8}$ as a single fraction in its simplest form.

8 Write $\dfrac{x^2 - x}{x^2 + x} \times \dfrac{x^2 + 2x + 1}{x - 1}$ as a single fraction in its simplest form.

C

B

A

A*

11.4 Algebraic proof

Objective

You can prove a given result using algebra.

Why do this?

Algebraic proofs are used to prove many key ideas (theorems) in life, from thermodynamics to quantum mechanics.

Get Ready

1. n is an integer.

State whether each of the following must represent an even number, an odd number or either.

a $2n$ **b** $n + 1$ **c** $2n - 1$ **d** n^2 **e** $3n$

Key Points

- To demonstrate that a result is true, you can give details of a particular case. For example, to demonstrate that the sum of two odd numbers is even, you could choose any two odd numbers and show that their sum is even, such as $3 + 5 = 8$.

- To **prove** that a result is true, you must show that it will be true in all cases. For example, to prove that the sum of two odd numbers is even, you must choose two 'general' odd numbers and show that their sum will always be even. You could write
  $(2m - 1) + (2n - 1) = 2m + 2n - 2 = 2(m + n - 1)$ which is an even number as it is a multiple of 2.

- In algebraic **proof** you will find the following points helpful.
  Where n is an integer:

  - Consecutive integers can be written in the form $n, n + 1, n + 2, n + 3, \ldots$ In some cases it is more useful to write them in a slightly different form, for example $n - 2, n - 1, n, n + 1, n + 2, \ldots$

  - Any even number can be written in the form $2n$.

  - Consecutive even numbers can be written in the form $2n, 2n + 2, 2n + 4, \ldots$

  - Any odd number may be written in the form $2n - 1$ (alternatively any odd number may be written in various other forms, for example $2n + 1$).

  - Consecutive odd numbers can be written in the form $2n - 1, 2n + 1, 2n + 3, \ldots$

Example 15

a Show that $(2n - 1)^2 + (2n + 1)^2 = 8n^2 + 2$.

b Hence prove that the sum of the squares of any two consecutive odd numbers is even.

a $(2n - 1)^2 + (2n + 1)^2 = (2n - 1)(2n - 1) + (2n + 1)(2n + 1)$ ← Write out the expression.

$$= 4n^2 - 4n + 1 + 4n^2 + 4n + 1$$ ← Multiply out the brackets.

$$= 8n^2 + 2$$ ← Simplify the expression.

b Any odd number may be written in the form $2n - 1$, where n is an integer. The next odd number can therefore be written as $2n + 1$.

> Add 2 to $2n - 1$.

So the sum of the squares of any two consecutive odd numbers $= (2n - 1)^2 + (2n + 1)^2$.

> Write the sum of the squares of any two consecutive odd numbers algebraically.

$$= 8n^2 + 2$$

> Use the identity given in part **a**.

$$= 2(4n^2 + 1)$$

> Show that 2 is a factor of the expression.

As $8n^2 + 2$ is a multiple of 2 it must represent an even number.

> Explain why the expression must represent an even number.

So, the sum of the squares of any two consecutive odd numbers is always even.

> Complete your proof by stating the result.

Exercise 11E

1. Prove that the sum of any odd number and any even number is odd. **A03**

*2. Prove that half the sum of four consecutive numbers is odd. **A03**

*3. Prove that the sum of any three consecutive numbers is a multiple of 3. **A03**

4. **a** Prove that the product of any odd number and any even number is even.
   b Prove that the product of any two odd numbers is odd.
   c Prove that the product of any two even numbers is even. **A03**

*5. Prove that for any two numbers the product of their difference and their sum is equal to the difference of their squares. **A03**

*6. Prove that, if the difference of two numbers is 4, then the difference of their squares is a multiple of 8. **A03**

Chapter review

- **Algebraic fractions**, like numerical fractions, can often be simplified.
- To simplify an algebraic fraction, factorise the numerator and the denominator, then divide the numerator and denominator by any common factors.
- To add (or subtract) algebraic fractions we use a similar method to that used for adding and subtracting numerical fractions.
- If the denominator of the fractions are the same, add (or subtract) the numerators but do not change the denominator.
- To add (or subtract) algebraic fractions with different denominators, find a common denominator and write each fraction as an equivalent fraction with this denominator.

- To find the lowest common denominator of algebraic fractions, you may need to factorise the denominators first.
- To simplify your answers you may have to factorise the numerator.
- To multiply (or divide) algebraic fractions we use a similar method to that used for multiplying and dividing numerical fractions.
- To multiply fractions, multiply the numerators and multiply the denominators.

$$\frac{a}{b} \times \frac{c}{d} = \frac{ac}{bd}$$

- To divide fractions, multiply the first fraction by the reciprocal of the second.

$$\frac{a}{b} \div \frac{c}{d} = \frac{a}{b} \times \frac{d}{c} = \frac{ad}{bc}$$

- You may need to factorise the numerator and/or the denominator before you multiply or divide algebraic fractions.
- To demonstrate that a result is true, you can give details of a particular case.
- To **prove** that a result is true, you must show that it will be true in all cases.
- In algebraic **proof** you will find the following points helpful.

  Where n is an integer:

  - Consecutive integers can be written in the form $n, n + 1, n + 2, n + 3, \ldots$ In some cases it is more useful to write them in a slightly different form, for example $n - 2, n - 1, n, n + 1, n + 2, \ldots$
  - Any even number can be written in the form $2n$.
  - Consecutive even numbers can be written in the form $2n, 2n + 2, 2n + 4, \ldots$
  - Any odd number may be written in the form $2n - 1$ (alternatively any odd number may be written in various other forms, for example $2n + 1$).
  - Consecutive odd numbers can be written in the form $2n - 1, 2n + 1, 2n + 3, \ldots$

Review exercise

A

1 Here are the first 4 lines of a number pattern.

$$1 + 2 + 3 + 4 = (4 \times 3) - (2 \times 1)$$
$$2 + 3 + 4 + 5 = (5 \times 4) - (3 \times 2)$$
$$3 + 4 + 5 + 6 = (6 \times 5) - (4 \times 3)$$
$$4 + 5 + 6 + 7 = (7 \times 6) - (5 \times 4)$$

n is the first number in the nth line of the number pattern.

Show that the above number pattern is true for the four consecutive integers

$$n, (n + 1), (n + 2) \text{ and } (n + 3)$$

Nov 2007

A*

2 Simplify fully.

a $\dfrac{4x^3}{8x}$

b $\dfrac{2(x + 1)^4}{6(x + 1)^2}$

c $\dfrac{x^2 + 5x}{x + 5}$

d $\dfrac{x^2 + 7x + 6}{x^2 + 8x + 12}$

e $\dfrac{x^2 - 2x}{x^2 + x - 6}$

f $\dfrac{x^2 - 25}{x^2 + 10x + 25}$

g $\dfrac{x^2 - 2x + 1}{x^3 - x}$

h $\dfrac{2x^2 + 7x - 4}{6x^2 + x - 2}$

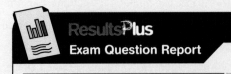

ResultsPlus
Exam Question Report

78% of students answered this sort of question poorly because they simplified at the wrong stage in the calculation.

A*

3 Write as a single fraction in its simplest form.

a $\dfrac{3x}{10} + \dfrac{x}{5}$

b $\dfrac{3}{2x} + \dfrac{2}{3x}$

c $\dfrac{1}{5x - 3} + \dfrac{1}{5x + 3}$

d $\dfrac{3}{(x + 1)(x + 2)} + \dfrac{2}{x + 1}$

e $\dfrac{1}{(x + 1)^2} + \dfrac{1}{x + 1}$

f $\dfrac{1}{x^2 + 4x + 3} + \dfrac{1}{x^2 + 8x + 15}$

4 Simplify.

a $\dfrac{x}{2} \times \dfrac{4}{x}$

b $\left(\dfrac{x}{3}\right)^2 \times \dfrac{9}{x}$

c $\dfrac{x}{10} \div \dfrac{x}{4}$

d $\dfrac{2s}{3} \div \dfrac{1}{s}$

e $\dfrac{x - 1}{(x + 1)^2} \times \dfrac{x + 1}{x + 3}$

f $\dfrac{x^2 - x}{x^2 + x} \times \dfrac{x^2 - 1}{(x - 1)^2}$

g $\dfrac{x + 2}{5} \div \dfrac{3x + 6}{10}$

h $\dfrac{2x - 1}{x^2} \div \dfrac{2x - 1}{2x^2 + x}$

5 a Show that $(n + 1)(n + 2) + n(n + 1) = 2(n + 1)^2$.

    b Hence prove that for any three consecutive integers, the sum of the product of the last two and the product of the first two is always even.

    A03

*6 Prove that the difference of the squares of two consecutive odd numbers is a multiple of 8.

    A03

7 Show that $25 - \dfrac{(x - 8)^2}{4} = \dfrac{(2 + x)(18 - x)}{4}$

June 2005   **A03**

8 The nth even number is $2n$.

    a Explain why the next even number after $2n$ is $2n + 2$.

    b Show algebraically that the sum of any 3 consecutive even numbers is always a multiple of 6.

    A03

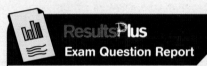

Results Plus
Exam Question Report

93% of students answered this question poorly because they did not follow the direction indicated in the early parts of the question.

June 2008, adapted

*9 Prove that $(3n + 1)^2 - (3n - 1)^2$ is a multiple of 4, for all positive integer values of n.

June 2009   **A03**

10 Jim runs 16 km from home at a speed of x kph. He then runs the same distance back home at a speed 1 kph slower.

Work out an expression, in terms of x, for the total time Jim took to run from home and back.

Give your answer as a single fraction in its simplest form.

    A02 A03

Humans are *bilateria* – animals whose right and left side are more or less symmetrical. Babies will spend more time looking at pictures of symmetrical faces rather than asysmmetrical ones, and if people are asked to rate faces for attractiveness, the most symmetrical ones often win. This preference is also seen in animals, with female swallows preferring males with more symmetrical tails and female zebra finches choosing the males with the most symmetrical coloured leg bands.

◉ Objectives

In this chapter you will:
- learn about line symmetry and rotational symmetry
- learn about some special types of 2D shapes
- look at both the metric and imperial measurement systems and learn how to convert units between and within these systems.

◐ Before you start

You need to:
- know how to measure and draw lines to the nearest mm, and how to measure and draw angles to the nearest degree
- be able to make sensible estimates for a range of measures
- know that measurements given to the nearest whole unit may be inaccurate by up to one half in either direction
- know metric and imperial units of length, mass and volume and have an idea of their relative sizes.

12.1 **Symmetry in 2D shapes**

⊙ Objectives

- ⊙ You can recognise line symmetry in 2D shapes.
- ⊙ You can draw lines of symmetry on 2D shapes.
- ⊙ You can recognise rotational symmetry in 2D shapes.
- ⊙ You can find the order of rotational symmetry of a 2D shape.
- ⊙ You can draw shapes with given line symmetry and/or rotational symmetry.

⊙ Why do this?

There are examples of 2D symmetry in the man-made and natural world, such as wheels, flowers and butterflies.

⊙ Get Ready

1. Trace this star.
Fold your tracing along the dotted line. What do you notice?
Place your tracing over the star and turn the tracing paper clockwise.
Keep turning the tracing paper until you get back to the starting position.
What do you notice?

🔍 Key Points

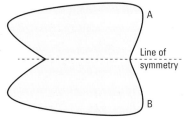

- ⊙ A shape has **line symmetry** if it can be folded so that one part of the shape fits exactly on top of the other part.
- ⊙ Every point of the shape on one side of the **line of symmetry** has a corresponding point on the **mirror image** the other side of the line. Notice that point A and its corresponding point B are the same distance from the line of symmetry.
- ⊙ If a mirror were placed on the line of symmetry, the shape would look the same. This is why line symmetry is sometimes called **reflection symmetry** and the line of symmetry is sometimes called the **mirror line**.
- ⊙ A shape has **rotation symmetry** if a tracing of the shape fits exactly on top of the shape in more than one position when it is rotated.
- ⊙ A tracing of a shape with rotation symmetry will fit exactly on top of the shape when turned through less than a complete turn.
- ⊙ The number of times that the tracing fits exactly on top of the shape is called the **order of rotational symmetry**.
- ⊙ Some two-dimensional shapes do not have any symmetry.

🔍 Example 1

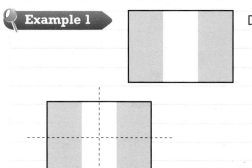

Draw in all the lines of symmetry on this flag.

> **ResultsPlus**
> **Examiner's Tip**
> 'All the lines' suggests that there is more than one line of symmetry.

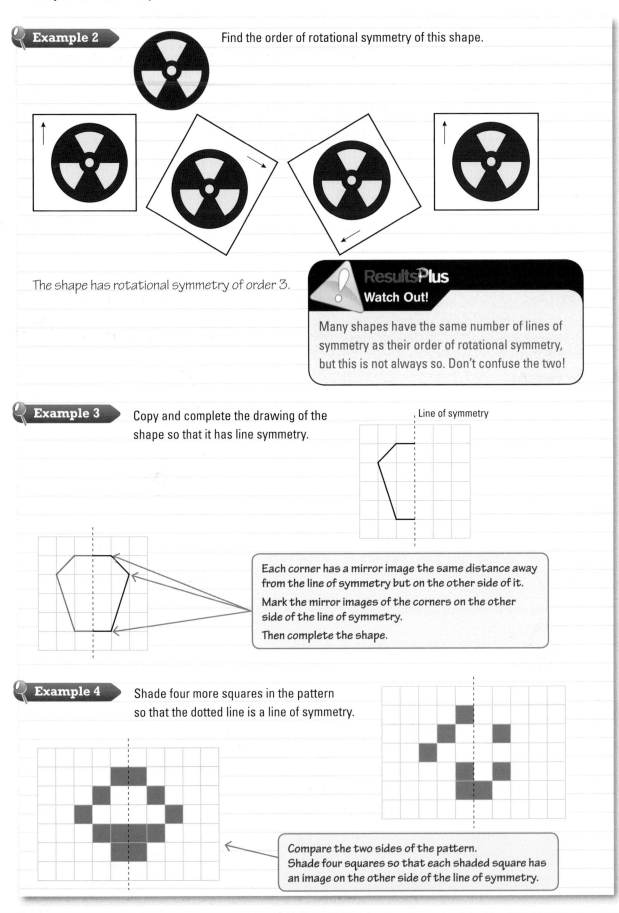

Example 2

Find the order of rotational symmetry of this shape.

The shape has rotational symmetry of order 3.

ResultsPlus

Watch Out!

Many shapes have the same number of lines of symmetry as their order of rotational symmetry, but this is not always so. Don't confuse the two!

Example 3

Copy and complete the drawing of the shape so that it has line symmetry.

Line of symmetry

Each corner has a mirror image the same distance away from the line of symmetry but on the other side of it.

Mark the mirror images of the corners on the other side of the line of symmetry.

Then complete the shape.

Example 4

Shade four more squares in the pattern so that the dotted line is a line of symmetry.

Compare the two sides of the pattern.
Shade four squares so that each shaded square has an image on the other side of the line of symmetry.

 **Example 5** The diagram shows part of a shape.

Complete the shape so that it has no lines of symmetry and rotational symmetry of order 2.

ResultsPlus
Examiner's Tip

Rotational symmetry of order 2 means that the completed shape must look the same when it is rotated through a half turn.

Exercise 12A

Questions in this chapter are targeted at the grades indicated.

1 For each shape, write down if it has line symmetry or not. If it has symmetry, copy the diagram and draw in all the lines.

a   b    c

d   e   f

2 Using tracing paper if necessary, state which of the following shapes have rotational symmetry and which do not have rotational symmetry.

For the shapes that have rotational symmetry, write down the order of the rotational symmetry.

a   b   c

d   e   f

3 a Copy and complete this shape so that it has line symmetry.

b Write down the name of the complete shape.

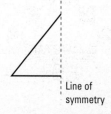

Line of symmetry

4   Each diagram shows an incomplete pattern.
For part **a**, copy the diagram and shade six more squares so that both dotted lines are lines of symmetry of the complete pattern. For part **b**, shade three more squares so that the complete pattern has rotational symmetry of order 4.

a

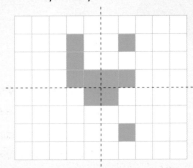

b

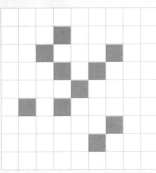

5   **a**   Draw a shape that has two lines of symmetry and rotational symmetry of order 2.
   b   Draw a shape with one line of symmetry and no rotational symmetry.
   c   Draw a shape that has no lines of symmetry and rotational symmetry of order 4.

12.2 Symmetry of special shapes

Objectives

- You know the symmetries of special triangles.
- You can recognise and name special quadrilaterals.
- You know the properties of special quadrilaterals.
- You know the symmetries of special quadrilaterals.
- You know the symmetries of regular polygons.

Why do this?

Many architectural designs are symmetrical in some way. The Taj Mahal, the Pyramids and the Greek Parthenon have impressive and beautiful uses of symmetry.

Get Ready

1. **a**   What is   **i**   an isosceles triangle   **ii**   an equilateral triangle?
   b   Is an equilateral triangle an isosceles triangle?
2. What is a quadrilateral?

Key Points

- **Triangles**
  A **triangle** is a polygon with three sides.
  Here is an isosceles triangle. It has two sides the same length.
  An isosceles triangle has one line of symmetry.

  Here is an equilateral triangle. All its sides are the same length.
  An equilateral triangle has three lines of symmetry and rotational symmetry of order 3.

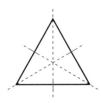

Quadrilaterals

A **quadrilateral** is a polygon with four sides. Some quadrilaterals have special names.
Here are some of the properties of special quadrilaterals.

Square

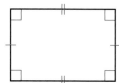

All sides equal in length.
All angles are 90°.
4 lines of symmetry and rotational symmetry of order 4.

Rectangle

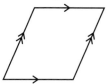

Opposite sides equal in length.
All angles are 90°.

Rhombus

All sides equal in length.
Opposite sides parallel.
Opposite angles equal.
2 lines of symmetry and rotational symmetry of order 4.

Parallelogram

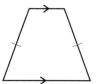

Opposite sides equal in length and parallel.
Opposite angles equal.
No lines of symmetry and rotational symmetry of order 2.

Trapezium

One pair of parallel sides.
No lines of symmetry and no rotational symmetry.

Isosceles trapezium

One pair of parallel sides.
Non-parallel sides equal in length.
One line of symmetry and no rotational symmetry.

Kite

Two pairs of **adjacent** sides equal in length.
(Adjacent means 'next to'.)
One line of symmetry and no rotational symmetry.

✹ Exercise 12B

1 **a** On squared paper, draw a right-angled triangle that has one line of symmetry.
Draw the line of symmetry on your triangle.
   b Write down what is special about this right-angled triangle.

2 Janine says, 'I am thinking of a quadrilateral. It has opposite sides that are parallel.'
   a Is there enough information to know what the quadrilateral is? Give reasons for your answer.
Janine now says, 'It has rotational symmetry of order 2.'
   b Is there now enough information to know what the quadrilateral is? Give reasons for your answer.
Janine now says, 'It has two lines of symmetry.'
   c Is there now enough information to know what the quadrilateral is? Give reasons for your answer.
Janine now says, 'It has sides that are not all the same length.'
   d What quadrilateral is Janine thinking of?

3 Draw a non-regular polygon which has a line of symmetry.

4 Draw a non-regular polygon which has rotational symmetry.
State the order of rotational symmetry of your polygon.

---

rectangle   rhombus   parallelogram   trapezium   isosceles trapezium   kite   adjacent   **145**

12.3 Converting between units of measure

◎ Objectives

○ You know the relationship between metric units and are able to convert between units in the metric system.

○ You know the approximate metric equivalents of imperial units and are able to convert between metric units and imperial units.

○ You are able to convert between units in the imperial system when you are given the relationship between the imperial units.

❓ Why do this?

It is important to be able to change from one unit to another when you are cooking as some measurements may be given in grams and some in kilograms.

⊕ Get Ready

1. The following sentences do not make sense because the wrong unit has been used. Rewrite each sentence using the correct unit.

   a The weight of a packet of biscuits is 150 kg.

   b The thickness of a book is 3 m.

   c The height of a giraffe is 5 km.

   d A teacup can hold 300 *l* of tea.

🔍 Key Points

◉ To **convert** from one metric unit to another metric unit it is necessary to know the following facts.

Length	Weight	Capacity/Volume
10 mm = 1 cm	1000 mg = 1 g	100 c*l* = 1 litre
100 cm = 1 m	1000 g = 1 kg	1000 m*l* = 1 litre
1000 mm = 1 m	1000 kg = 1 tonne	1000 cm³ = 1 litre
1000 m = 1 km		1000 *l* = 1 m³

You then only need to multiply or divide by 10, 100 or 1000 in order to convert between the different metric units.

◉ When you convert from a smaller unit to a larger unit, you need to divide.

◉ When you convert from a larger unit to a smaller unit, you need to multiply.

For example, for lengths:

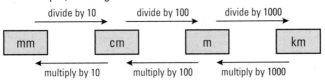

◉ To convert between a metric unit and an imperial unit it is necessary to know the facts in this table.

ResultsPlus

Watch Out!

The values in the table are not exact but they are the rough equivalents that need to be used in examinations.

Metric unit	Imperial unit
1 kg	2.2 pounds
1 litre (*l*)	$1\frac{3}{4}$ pints = 1.75 pints
4.5 *l*	1 gallon
8 km	5 miles
30 cm	1 foot
2.54 cm	1 inch

Example 6   a Convert 12 m into centimetres.
              b Convert 2670 g into kilograms.

a $12\,m = 12 \times 100\,cm = 1200\,cm$

> Centimetres are smaller than metres so there are more of them. As $100\,cm = 1\,m$, multiply by 100.

wrong

b $2670\,g = 2760 \div 1000 = 2.76\,kg$ $2.67\,kg$

> Kilograms are larger than grams so there are fewer of them. As $1000\,g = 1\,kg$, divide by 1000.

Exercise 12C

1 Convert these lengths to centimetres.

   a 6 m   b 210 mm   c 5.1 m   d 0.84 m
   e 59 mm   f 483 mm   g 3 km   h 0.067 km

2 Convert these weights to kilograms.

   a 3 tonnes   b 8.2 tonnes   c 6000 g
   d 900 g   e 430 g   f 4700 g

3 Convert these volumes to litres.

   a 2000 ml b 700 cl c 5900 ml d 45 000 ml

4 A bottle of lemonade contains 70 cl.
   How many litres of lemonade are there in 10 of these bottles?

Example 7   a Convert 8 gallons into litres.   b Convert 28 km into miles.

a $8\,gallons = 8 \times 4.5\,litres = 36\,litres$

> There are more litres than gallons so multiply.

b $8\,km = 5\,miles$
    $1\,km = 5 \div 8\,miles$
    $28\,km = 28 \times 5 \div 8\,miles$
    $= 140 \div 8 = 17.5\,miles$

> There are fewer miles than km as a mile is longer than a km. Find 1 km. Multiply and then divide.

Exercise 12D

1 Convert 4 kg to pounds.

2 Convert 110 pounds to kilograms.

3 Convert 7 pints to litres.

4 Convert 36 litres to gallons.

5 Convert 10 litres to pints.

6 Convert 12 feet to centimetres.

7 Convert 96 km to miles.

8 Convert 60 miles to kilometres.

A02
A03

9 The price of petrol is 130p per litre.
Work out the price of the petrol per gallon.

Example 8 ▶ There are 12 inches in a foot.
    a Convert 7 feet into inches.
    b Convert 108 inches into feet.

a 7 feet = 7 × 12 = 84 inches ◀

> A foot is longer than an inch so there are more inches than feet. So to change from feet to inches you multiply.

ResultsPlus
Examiner's Tip

When converting between imperial units you will not be expected to know the relationship between the units as the conversions will be given.

b 108 inches = 108 ÷ 12 = 9 feet

> To convert from inches to feet you divide.

Exercise 12E

1 There are 16 ounces in a pound.
    a Convert 5 pounds to ounces.
    b Convert 96 ounces to pounds.

A02

2 There are 12 inches in a foot and 3 feet in a yard.
Work out the number of inches in 5 yards.

A02

3 There are 14 pounds in a stone. John's weight is 9 stones and 6 pounds.
    a Work out John's weight in pounds.
    b Work out John's weight in kilograms.

Chapter review

◉ A shape has **line symmetry** if it can be folded so that one part of the shape fits exactly on top of the other part.

◉ Every point of the shape on one side of the **line of symmetry** has a corresponding point on the **mirror image** the other side of the line. Notice that corresponding points are the same distance from the line of symmetry.

◉ If a mirror were placed on the line of symmetry of a shape, the shape would look the same in the mirror. This is why line symmetry is sometimes called **reflection symmetry** and the line of symmetry is sometimes called the **mirror line**.

◉ A shape has **rotation symmetry** if a tracing of the shape fits exactly on top of the shape in more than one position when it is rotated.

◉ A tracing of a shape with rotation symmetry will fit exactly on top of the shape when turned through less than a complete turn.

◉ The number of times that the tracing fits exactly on top of the shape is called the **order of rotational symmetry**.

◉ Some two-dimensional shapes do not have any symmetry.

◉ **Triangles**

A **triangle** is a polygon with three sides.

Here is an isosceles triangle. It has two sides the same length.

An isosceles triangle has one line of symmetry.

Here is an equilateral triangle. All its sides are the same length.

An equilateral triangle has three lines of symmetry and rotational symmetry of order 3.

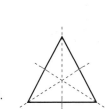
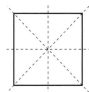

◉ **Quadrilaterals**

A **quadrilateral** is a polygon with four sides. Some quadrilaterals have special names.

Here are some of the properties of special quadrilaterals.

Square

All sides equal in length.
All angles are 90°.
4 lines of symmetry and rotational symmetry of order 4.

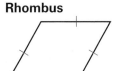

Rectangle

Opposite sides equal in length.
All angles are 90°.

Rhombus

All sides equal in length.
Opposite sides parallel.
Opposite angles equal.
2 lines of symmetry and rotational symmetry of order 4.

Parallelogram

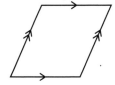

Opposite sides equal in length and parallel.
Opposite angles equal.
No lines of symmetry and rotational symmetry of order 2.

Trapezium

One pair of parallel sides.
No lines of symmetry and no rotational symmetry.

Isosceles trapezium

One pair of parallel sides.
Non-parallel sides equal in length.
One line of symmetry and no rotational symmetry.

Kite

Two pairs of **adjacent** sides equal in length.
(Adjacent means 'next to'.)
One line of symmetry and no rotational symmetry.

To **convert** from one metric unit to another metric unit it is necessary to know the following facts.

Length	Weight	Capacity/Volume
10 mm = 1 cm	1000 mg = 1 g	100 cl = 1 litre
100 cm = 1 m	1000 g = 1 kg	1000 ml = 1 litre
1000 mm = 1 m	1000 kg = 1 tonne	1000 cm³ = 1 litre
1000 m = 1 km		1000 l = 1 m³

You then only need to multiply or divide by 10, 100 or 1000 in order to convert between the different metric units.

When you convert from a smaller unit to a larger unit, you need to divide.

When you convert from a larger unit to a smaller unit, you need to multiply.

To convert between a metric unit and an imperial unit it is necessary to know the facts in the following table.

Metric unit	Imperial unit
1 kg	2.2 pounds
1 litre (l)	$1\frac{3}{4}$ pints = 1.75 pints
4.5 l	1 gallon
8 km	5 miles
30 cm	1 foot
2.54 cm	1 inch

Review exercise

1　**a** On the diagram below, shade **one** square so that the shape has exactly **one** line of symmetry.

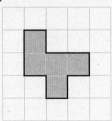

b On the diagram below, shade **one** square so that the shape has rotational symmetry of order 2.

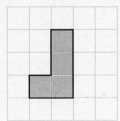

Nov 2008

2　**a** Complete the table by writing a sensible **metric** unit for each measurement.

The length of the river Nile	6700 kilometres
The height of the world's tallest tree	110 ....
The weight of a chicken's egg	70 ....
The amount of petrol in a full petrol tank of a car	40 ....

b Convert 4 metres to centimetres.

c Convert 1500 grams to kilograms.

June 2008

A03

3 Shalim says 1.5 km is less than 1400 m.

Is he right?

Explain your answer.

June 2007

4 **a** Write down a sensible **metric** unit that can be used to measure:

    i the height of a tree   **ii** the weight of a person.

  b Change 2 centimetres to millimetres.

Nov 2008

5 **a** Write down a sensible **metric** unit for measuring:

    i the distance from London to Paris

    ii the amount of water in a swimming pool

  b **i** Change 5 centimetres to millimetres

    ii Change 4000 grams to kilograms.

Nov 2008

6 Here is a biohazard sign.

  a How many lines of symmetry has this sign?

  b What is the order of rotational symmetry of this sign?

  c John has to fix this sign on a wall.

    All he knows is that the sign has to be fixed with a corner pointing upwards.

    How does the symmetry of the shape help John?

7 Here are four shapes.

A

B

C

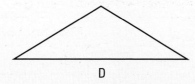

D

Write down the letter of the shape which has:

a exactly **one** line of symmetry   **b** **no** lines of symmetry   **c** exactly **two** lines of symmetry.

Nov 2008

8   Here are five shapes.

A

B

C

D

E

Two of these shapes have only **one** line of symmetry.

a   Write down the letter of each of these **two** shapes.

Two of these shapes have rotational symmetry of order 2.

b   Write down the letter of each of these **two** shapes.

June 2007

9   The shape below has one line of symmetry.   The shape below has rotational symmetry.

a   On the grid, draw this line of symmetry.   b   Write down the order of rotational symmetry.

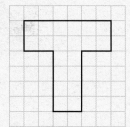

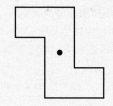

Nov 2007

10   Carla has a bottle of water. There are 2 litres of water, correct to the nearest litre, in Carla's bottle.
Gurpia has a jug of water. There are 1950 millilitres of water, correct to the nearest millilitre, in Gurpia's jug.
Carla says that she has more water than Gurpia.
Is Carla necessarily right? Explain your answer.

A curling bridge looks like a conventional bridge when it is extended. However, it curls up to form an octagon to allow boats through. This Rolling Bridge is in Paddington Basin in London, and curls up every Friday at midday.

Objectives

In this chapter you will:

- recognise and use corresponding angles and alternate angles
- use and prove angle properties of triangles and quadrilaterals
- recognise angles of elevation and depression
- give reasons for angle calculations
- recognise and know the names of special polygons
- know and use the interior and exterior angle properties of polygons
- learn about the geometric properties of circles and tangents.

Before you start

You should know:

- how to find the sizes of missing angles on a straight line and at the intersection of straight lines
- how to recognise perpendicular and parallel lines, and vertically opposite angles
- that the angle sum of a triangle is 180°
- how to recognise scalene, isosceles, equilateral and right-angled triangles, and use their angle properties
- how to recognise acute, obtuse, reflex and right angles
- that a quadrilateral is a shape with four straight sides and four angles
- what a circle, semicircle and quarter circle are, and be able to name the parts of a circle and related terms.

13.1 Angle properties of parallel lines

Objectives

- You can mark parallel lines on a diagram.
- You can recognise corresponding and alternate angles.
- You can find the sizes of missing angles using corresponding and alternate angles.
- You can give reasons for angle calculations.

Get Ready

1. Find the sizes of the marked angles.

a

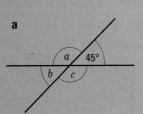

b

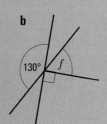

c

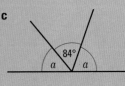

Key Points

- When two **parallel** lines are crossed by a straight line, as in the diagram, angles are formed.

  The green angles are in corresponding positions so they are called **corresponding angles**. Corresponding angles are equal.

  The blue angles are also corresponding angles.

  The orange angles are on opposite or alternate sides of the line so they are called **alternate angles**.

  Alternate angles are equal.

  The yellow angles are also alternate angles.

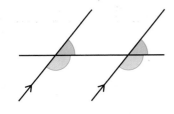

Example 1

Find the size of angle a and angle b.

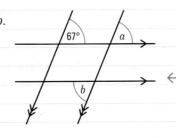

$a = 67°$

Corresponding angles are equal.

$b = a$

so

Alternate angles are equal.

$b = 67°$

Parallel lines are marked with arrows in diagrams.

parallel   corresponding angles   alternate angles

A02
A03

> **Example 2** ▶ Explain why $a + b = 180°$.

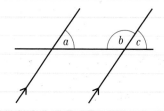

In the diagram
$a = c$
$b + c = 180°$
so
$b + a = a + b = 180°$

> Corresponding angles are equal.

> Angles on a straight line add up to 180°.

Exercise 13A

Questions in this chapter are targeted at the grades indicated.

In questions 1–6 find the size of each lettered angle. Give reasons for your answers.

1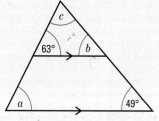

2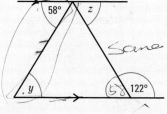

3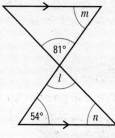

4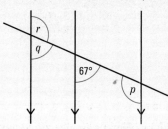

5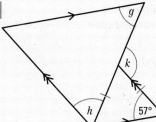

6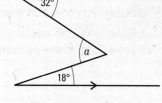

7 Here are two parallel lines crossed by a straight line.
   a List pairs of equal corresponding angles.
   b List pairs of equal alternate angles.
   c List pairs of angles which add up to 180°.
     Explain why the angles add up to 180°.

D

8 ACE is a straight line.
Explain why the lines AB and CD must be parallel.

A03

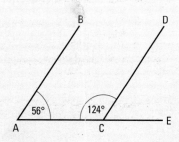

155

13.2 Proving the angle properties of triangles and quadrilaterals

◎ Objectives

- ◉ You can understand a proof that the angle sum of a triangle is 180°.
- ◉ You can understand a proof that an exterior angle of a triangle is equal to the sum of the interior angles at the other two vertices.
- ◉ You can explain why the angle sum of a quadrilateral is 360°.

⑦ Why do this?

In mathematics it is important to be able to prove that results are always true. A demonstration only shows that the result is true for the chosen values.

⬆ Get Ready

In questions 1 – 3, calculate the size of each lettered angle. Give reasons for your answers.

1.
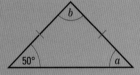

2.
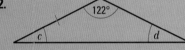

3.

◉ Key Points

- ◉ You can use angle facts to prove angle properties.
  - ◉ The angle marked e is called an **exterior angle**.
  - ◉ The angle of the triangle at this vertex, i, is sometimes called an **interior angle**.
  - ◉ $i + e = 180°$

📌 Example 3

A03

Prove that the angle sum of any triangle is 180°.

For any triangle, a straight line can be drawn through a vertex parallel to the opposite side, as shown in the diagram.

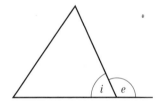

$a = p$ ← Alternate angles are equal.

$c = q$ ← Alternate angles are equal.

$p + b + q = 180°$ ← Angles on a straight line add up to 180°.

So
$a + b + c = 180°$ ← The angle sum of a triangle is 180°.

Exercise 13B

1   Here is a triangle with one side extended.

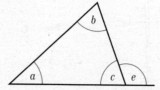

Complete the following proof that $e = a + b$ by giving a reason for each line of the proof.

$a + b + c = 180°$ ....

$e + c = 180°$ ....

so

$e = a + b$

An exterior angle of a triangle is equal to the sum of the interior
angles at the other two vertices.

2   Here is a quadrilateral. A diagonal has been drawn to divide the quadrilateral into two triangles.

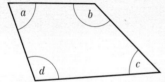 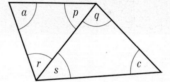

Copy and complete this proof that the angle sum of a quadrilateral is 360°.

$b = p + q$ ....

$d = r + s$ ....

$a + p + r = 180°$ ....

$c + q + s = 180°$ ....

Adding

$a + c + p + q + r + s = 360°$

$a + c + b + d = 360°$

so

$a + b + c + d = 360°$

The sum of the angles of a quadrilateral is 360°.

3

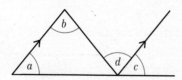

a   Use properties of parallel lines to prove that

$a + b = c + d$

b   Which angle property of triangles has this proved?

13.3 Using the angle properties of triangles and quadrilaterals

◉ Objectives

○ You can use the property that an exterior angle of a triangle is equal to the sum of the interior angles at the other two vertices.

○ You can use the property that the angle sum of a quadrilateral is 360°.

○ You can use the angle properties of a parallelogram.

○ You can give reasons for angle calculations.

? Why do this?

The angles in triangles are used in sports, for example in water-skiing. To ensure the longest jumps are made, the angle of the jump should be 14° to the water.

↟ Get Ready

1. Calculate the size of the angle marked j.

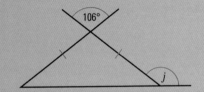

Key Points

◉ An exterior angle of a triangle is equal to the sum of the interior angles at the other two vertices.

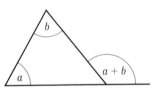

◉ The angle sum of a quadrilateral is 360°.
$a + b + c + d = 360°$

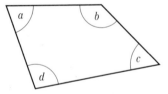

◉ Opposite angles of a parallelogram are equal. $a = c$
$b = d$

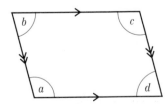

The two angles at the end of each side of a parallelogram add up to 180°.
$a + b = b + c = c + d = d + a = 180°$

Example 5 shows important angle properties of parallelograms.

Example 4

BCD is a straight line.
angle ACD = 123°
angle CBA = 58°
Work out the size of angle BAC.

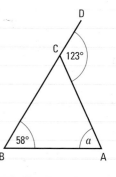

$a + 58° = 123°$ ← Angle DCA is an exterior angle of the triangle.

$a = 123° - 58°$ ← Exterior angle is equal to the sum of the interior angles at the other two vertices.

$a =$ angle BAC $= 65°$

Example 5

ABCD is a parallelogram.
Find the size of each angle of the parallelogram.

$a = 60°$ ← Alternate angles on parallel lines BC and AD are equal.

$b + 60° = 180°$
$b = 180° - 60°$ ← Angles on a straight line add up to 180°.

$b = 120°$

$c = 60°$ ← Corresponding angles on parallel lines BA and CD are equal.

$a + b + c + d = 360°$ ← Angle sum of a quadrilateral is 360°.
$60° + 120° + 60° + d = 360°$
$240° + d = 360°$
$d = 120°$

Exercise 13C

1 BCDE is a quadrilateral.
ABE is an equilateral triangle.
Work out the size of angle ABC.

2 Work out the size of the angle e.
Give reasons for your working.

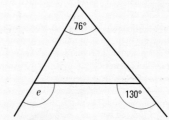

D

3 **a** Here is a kite. The diagonal shown dotted is an axis of symmetry.
Find the size of angle a and the size of angle b.
Give reasons for your working.

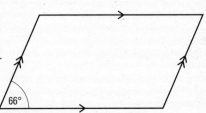

b Here is a parallelogram.
The parallelogram has an angle of 66° as shown.
Find the sizes of the three other angles of the parallelogram.
Give reasons for your working.

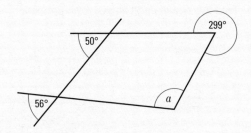

A03

4 Here is a quadrilateral.
Work out the size of angle a.
Give reasons for your working.

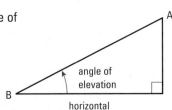

13.4 Angles of elevation and depression

Objectives

○ You can recognise an angle of elevation.
○ You can recognise an angle of depression.
○ You know that the angle of elevation of point A
from point B is equal to the angle of depression
of point B from point A.

Why do this?

If you wanted to abseil down a building, you could
work out the height of the building using the angle
of elevation from a point on the ground to the top.

Get Ready

1.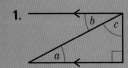
    a Explain why $a = b$.
    b What is $b + c$?

Key Points

● The **angle of elevation** of point A from point B is the angle of
turn above the horizontal to look directly from B to A.

● The **angle of depression** of point B from point A is the angle of
turn below the horizontal to look directly from A to B.

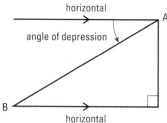

● Angles of elevation and depression are always measured from the horizontal.

Exercise 13D

1 Here is a diagram of a lighthouse.
Explain why angle e = angle d.

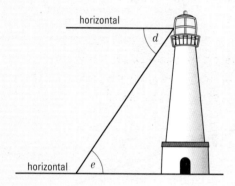

2 The angle of depression of a point A on horizontal ground from the top of a tree is 30°.
  a Show this information in a sketch.
  b On your sketch show and label the angle of elevation of the top of the tree from A.

13.5 Using angle properties to solve problems

◎ **Objectives**

● You can use the angle properties in this chapter
to solve more involved problems.
● You can give reasons for angle calculations.

❓ **Why do this?**

In the fashion industry, you can use angle
properties to fit pieces of material together for
clothes without wasting cloth.

⬆ **Get Ready**

1. Find the size of each lettered angle.
Give reasons for your answers.

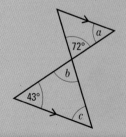
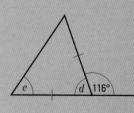
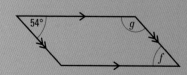

Example 6 Work out the size of:

i angle a

ii angle b.

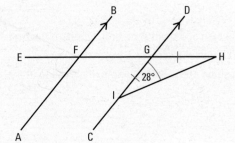

i $a + 75° + 120° + 103° = 360°$

$a + 298° = 360°$

$a = 62°$

Sum of the angles of a quadrilateral is 360°.

ii $b = a$

$b = 62°$

Alternate angles are equal.

ResultsPlus

Watch Out!

Questions will rarely just ask you to work out the size of the unknown angle of a triangle or of a quadrilateral. In most cases you will need to use other angle properties.

Exercise 13E

D
A02
A03

1 AFB and CIGD are parallel lines.

EFGH is a straight line.

GH = GI

Angle GIH = 28°

Work out the size of:

a angle DGF

b angle EFA.

C

2 L, M and N are points, as shown, on the sides of triangle ABC.

ML and AB are parallel.

NL and AC are parallel.

NM and BC are parallel.

Angle BAC = 70°

Angle ABC = 55°

Work out the size of each angle of triangle LMN.

A02
A03

3 Work out the size of:

a angle p

b angle q.

Give reasons for your working.

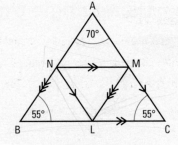

A02
A03

4 Here is a parallelogram.

a Explain why $a = c$.

b Hence prove that $a + b = c + d$.

Give reasons for your working.

c What property of parallelograms have you proved?

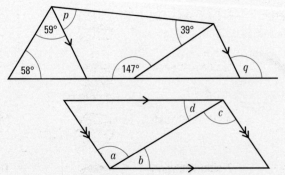

162

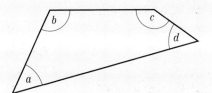

5 Here is a quadrilateral.
In this quadrilateral $a + c = 180°$.
Prove that $b + d = 180°$.
Give reasons for your working.

13.6 Polygons

⊙ Objectives

- ◉ You know what a polygon is and what a regular polygon is.
- ◉ You know the names of special polygons.
- ◉ You know and can use the sum of the interior angles of a polygon.
- ◉ You know and can use the sum of the exterior angles of a polygon.
- ◉ You can answer problems on polygons involving angles.

⟳ Why do this?

Tiles are made in the shape of regular polygons. Knowing the properties of various polygons might help you make patterns with different shapes.

⬥ Get Ready

1. Write down the name of the shape that has three equal sides and three equal angles.
2. Use two words to complete the following sentence: A square is a quadrilateral with... sides and... angles.

Key Points

- ◉ A **polygon** is a closed two-dimensional shape with straight sides.
- ◉ A **regular** polygon is a polygon with all its sides the same length and all its angles equal in size.
- ◉ Here are the names of some special polygons.

Triangle	3-sided polygon
Quadrilateral	4-sided polygon
Pentagon	5-sided polygon
Hexagon	6-sided polygon
Heptagon	7-sided polygon

| Octagon | 8-sided polygon |
| Decagon | 10-sided polygon |

Results Plus
Examiner's Tip

These are the names of the polygons that are needed for GCSE. Other polygons also have special names.

Angles of a polygon

Example 7 Here is a 7-sided polygon.

Work out the sum of the angles of this polygon.

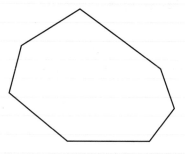

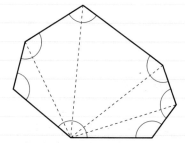

A 7-sided polygon is a heptagon.

All the diagonals from one vertex (corner) of the heptagon have been drawn.

There are 4 diagonals and the heptagon has been divided into 5 triangles.

The angle sum of each triangle is 180°.

Sum of the angles of a heptagon = 5 × 180° = 900°

As these angles are inside the polygon, they are also called the interior angles of the polygon.

Exercise 13F

1 **a** Copy and complete the following table.

Polygon	Number of sides (n)	Number of diagonals from one vertex	Number of triangles formed	Sum of interior angles
Triangle	3	0	1	180°
Quadrilateral	4	1	2	360°
Pentagon	5			
Hexagon	6			
Heptagon	7	4	5	900°
Octagon	8			
Nonagon	9			
Decagon	10			

b For a polygon with n sides, write down:
  i the number of diagonals that can be drawn from one vertex
  ii the number of triangles that are formed
  iii the sum of the interior angles of the polygon.

2 A rhombus has sides that are the same length. Explain why a rhombus is, in general, not a regular polygon.

Sum of the interior angles of a polygon

Key Points

⊙ A polygon can be divided into triangles when all diagonals are drawn from one vertex. For an n-sided polygon, the number of triangles will be $(n - 2)$.

⊙ A regular polygon will tesselate if the interior angle is an exact divisor of 360°. Regular triangles, squares and hexagons will therefore tessellate.

⊙ Sum of the interior angles of a polygon with n sides $= (n - 2) \times 180°$
$$= (2n - 4) \text{ right angles}$$

Example 8

A polygon has 15 sides.

a Work out the sum of the interior angles of the polygon.

b Find the size of each interior angle of a regular polygon with 15 sides.

a
$$15 - 2 = 13 \quad \longleftarrow \quad \boxed{\text{With } n = 15, \text{ work out the number of triangles} = (n - 2)}$$

$$13 \times 180 = 2340 \quad \longleftarrow \quad \boxed{\text{Work out } (n - 2) \times 180}$$
$$\text{Sum of interior angles} = 2340°$$

b Each interior angle $= 2340° \div 15 \quad \longleftarrow \quad \boxed{\text{The regular polygon has 15 interior angles that are all the same size.}}$

$$= 156° \quad \longleftarrow \quad \boxed{\text{Divide the sum of the angles by 15.}}$$

Example 9

Here is a regular octagon with centre O.

a Work out the size of:

    i angle x ii angle y.

b Hence work out the size of each interior angle of a regular octagon.

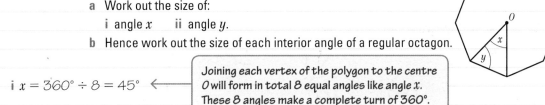

a i $x = 360° \div 8 = 45° \quad \longleftarrow \quad \boxed{\text{Joining each vertex of the polygon to the centre } O \text{ will form in total 8 equal angles like angle } x. \text{ These 8 angles make a complete turn of 360°.}}$

  ii $y = \dfrac{180° - 45°}{2} = 67.5° \quad \longleftarrow \quad \boxed{\text{The triangle shown is an isosceles triangle, with angle } y \text{ as one of the two equal base angles.}}$

b Each interior angle $= 2 \times 67.5° = 135° \quad \longleftarrow \quad \boxed{\text{By symmetry, angle } y \text{ is half an interior angle of the regular polygon.}}$

Exercise 13G

1 John divides a regular polygon into 16 triangles by drawing all the diagonals from one vertex.

    a How many diagonals does John draw?

    b How many sides has the polygon?

    c What is the size of each of the interior angles of the polygon?

A02
A03

D

D AO2 AO3

2 Work out the size of each interior angle of:
a a regular hexagon
b a regular decagon
c a regular polygon with 30 sides.

C AO3

3 Work out the size of each of the marked angles in these polygons. You must show your working.

a

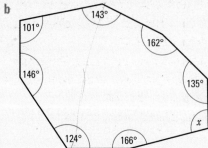

b

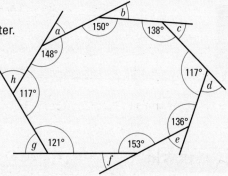

AO2 AO3

4 Explain why the size of the angle at the centre of a regular polygon cannot be 25°.

5 Here is an octagon.
a Work out the size of each of the angles marked with a letter.
b Work out the value of $a + b + c + d + e + f + g + h$

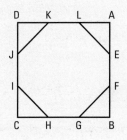

B AO2 AO3

6 ABCD is a square. EFGHIJKL is an octagon.
AE = EF = FB = BG = GH = HC = CI = IJ = JD = DK = KL = LA
a Find the size of each interior angle of the octagon.
   Give reasons for your answers.
b Tracy says that the octagon is a regular octagon.
   i Why might Tracy think that the octagon is regular?
   ii Explain why Tracy is wrong.

Exterior angles of a polygon

Key Points

◉ When a side of a polygon is extended at a vertex, the angle between this extended line and the other side at the vertex is an exterior angle at this vertex.

interior angle

exterior angle

◉ The sum of angles on a straight line = 180°
So at a vertex, interior angle + exterior angle = 180°

◉ The sum of the exterior angles of any polygon is 360°.

ABCDEF is a hexagon.

Imagine a spider is at vertex A facing in the direction of the arrow.
The spider turns through angle a so that it is now facing in the
direction AB. The spider now walks to vertex B.

At B, the spider turns through angle b to face in the direction BC.
He continues to walk around the hexagon until he gets back to A.
The spider has turned through one complete circle, so it has
turned through an angle of 360°.

The total angle turned through by the spider is also
$a + b + c + d + e + f$, the sum of the exterior angles of the hexagon.
So $a + b + c + d + e + f = 360°$.

The same argument holds for any polygon.

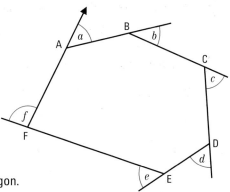

Example 10 ▶ A regular polygon has 20 sides.

    a Work out the size of each exterior angle.

    b Work out the size of each interior angle.

a Each exterior angle = 360° ÷ 20 = 18°   ← | The polygon is regular so the 20 exterior angles are equal in size.
The sum of these 20 equal angles is 360°.

b Interior angle = 180° − 18° = 162°   ← | Exterior angle + interior angle = 180°

Example 11 ▶ The interior angle of a regular polygon is 160°.

    Work out how many sides the polygon has.

Exterior angle = 180° − 160° = 20°   ← | Work out the size of an exterior angle.
Exterior angle + interior angle = 180°.

Number of sides = $\frac{360°}{20°}$ = 18   ← | The polygon is regular so that all exterior angles are 20° with sum 360°.

⚙ **Exercise 13H**

1 One vertex of a polygon is the point P.

    a Work out the size of the interior angle at P when the exterior angle at P is: **i** 70°   **ii** 37°.

    b Work out the size of the exterior angle at P when the interior angle at P is: **i** 130°   **ii** 144°.

2 Work out the size of each exterior angle of:

    a a regular pentagon   **b** a regular octagon

    c a regular polygon with 12 sides   **d** a regular 25-sided polygon.

3 The size of each exterior angle of a regular polygon is 15°.

    a Work out the number of sides the polygon has.

    b What is the sum of the interior angles of the polygon?

D

A02
A03

C

A03

4 The sizes of five of the exterior angles of a hexagon are 36°, 82°, 51°, 52° and 73°.
Work out the size of each of the interior angles of the hexagon.

A02
A03

5 A, B and C are three vertices of a regular polygon with 30 sides.

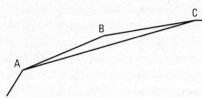

Work out the size of angle BCA.
Give reasons for your working.

A02
A03

* **6** The diagram shows three sides, AB, BC and CD, of a regular polygon with centre O.
The angle at the centre of the polygon is c.
The exterior angle of the polygon at the vertex C is e.

Explain why $c = e$.

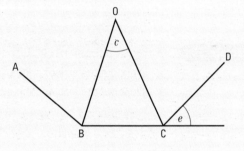

13.7 Isosceles triangle in a circle

◎ Objective

○ You can use the properties of angles in a circle.

❷ Why do this?

Ferris wheels are constructed using the properties of an isosceles triangle in a circle.

⬆ Get Ready

1. Calculate the size of the angles marked a, b and c.

a

b

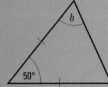

c

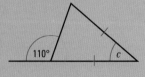

🔍 Key Point

◉ The radii of a circle are all the same length.
This means that a triangle in a circle where two of its sides are radii is an isosceles triangle.

Example 12

P and Q are points on the circumference of a circle, centre O.
PQR is a straight line.
Angle OQR = 150°
Calculate the size of angle POQ.
Give reasons for your answer.

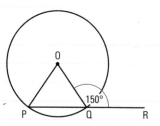

Angle OQP = 180° − 150°   ← | Adjacent angles on a straight line. |
   = 30°
Reason: The angles on a straight line add up to 180°.

OP = OQ   ← | Radii equal, thus isosceles triangle. |

Angle QPO = Angle OQP = 30°   ← | Mark the angles on the diagram. |

Reason: In an isosceles triangle the angles opposite
the equal sides (radii equal) are the same size.

Angle POQ = 180° − 30° − 30°   ← | Subtract the two angles you know from 180°. |
   = 120°
Reason: The angles in a triangle add up to 180°.

ResultsPlus
Examiner's Tip

Write down the reason; this is the rule that you have used and should be in words.

 Exercise 13I

The diagrams all show circles, centre O.
Work out the size of each angle marked with a letter.

1
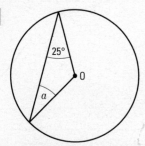

2
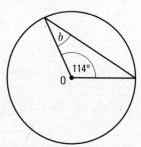

3
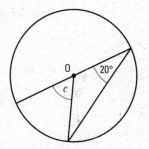

In questions 4–6 give reasons for your answers.

* 4
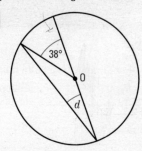

* 5
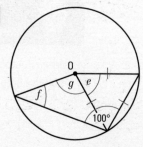

* 6
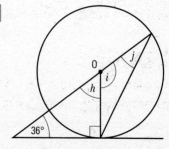

169

13.8 Tangents to a circle

⊙ **Objective**

○ You can use the properties of tangents to a circle.

⊙ **Why do this?**

You would need to understand tangents to a circle to design the gear system on a bike.

⬆ **Get Ready**

1. What are the values of angles a, b and c in this rectangle?

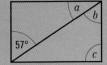

🔍 **Key Points**

◉ The angle between a tangent and a radius of a circle is 90°.
Angle OQP = angle OQR = 90°

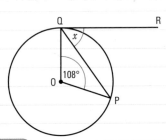

◉ Tangents to a circle from a point outside the circle are equal in length.
AB = BC

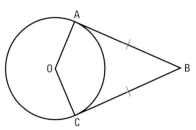

 A03

🔍 **Example 13**

QR is a tangent to the circle, centre O.
PQ is a chord of the circle.
Angle POQ = 108°
Work out the value of x.
Give reasons for your answer.

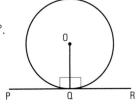

Angle OQP = (180° − 108°) ÷ 2   ← | Isosceles triangle, radii equal. |
= 72° ÷ 2
= 36°   ← | Put 36° on the diagram. |

OQP is an isosceles triangle as the radii are equal.   ← | Write all the different reasons in words. |
In an isosceles triangle the angles opposite the
equal sides are the same size and the angles in a
triangle add up to 180°.

💡 Results**Plus**
Examiner's Tip

When you have calculated an angle, you can mark
it on the diagram to help you answer the question.

x = 90° − angle OQP
= 90° − 36°
= 54°
The angle between the tangent and the radius is 90°.   ← | Write this reason too. |

Example 14   RP and RQ are tangents to the circle, centre O.

PQ is a chord.

Work out the value of y.

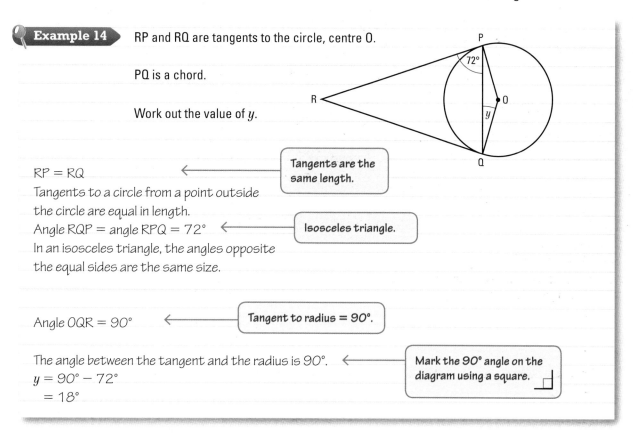

RP = RQ   ← Tangents are the same length.

Tangents to a circle from a point outside the circle are equal in length.

Angle RQP = angle RPQ = 72°   ← Isosceles triangle.

In an isosceles triangle, the angles opposite the equal sides are the same size.

Angle OQR = 90°   ← Tangent to radius = 90°.

The angle between the tangent and the radius is 90°.   ← Mark the 90° angle on the diagram using a square.

$y = 90° - 72°$

$= 18°$

Exercise 13J

The diagrams all show circles, centre O.
Work out the size of each angle marked with a letter.

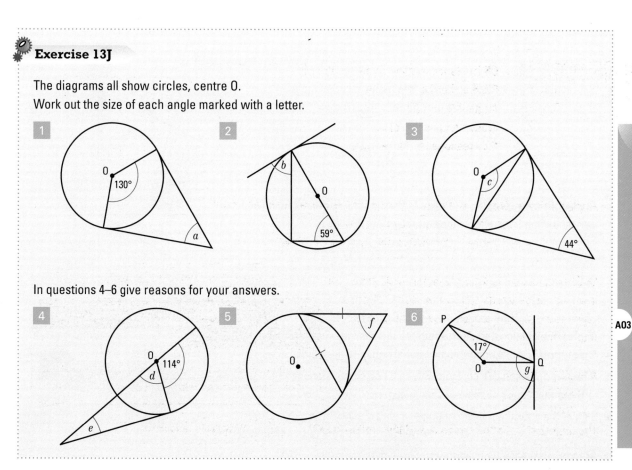

In questions 4–6 give reasons for your answers.

B

A03

Chapter review

- When two **parallel** lines are crossed by a straight line, as in the diagram, angles are formed. **Corresponding angles** are equal, and **alternate angles** are equal.
- You can use angle facts to prove angle properties.
  - The angle marked e is called an **exterior angle**.
  - The angle of the triangle at this vertex, i, is sometimes called the **interior angle**.
  - $i + e = 180°$

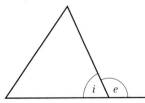

- The angle sum of a triangle is 180°.
- An **exterior angle** of a triangle is equal to the sum of the **interior angles** at the other two vertices.

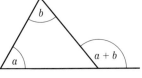

- The angle sum of a quadrilateral is 360°.
  $a + b + c + d = 360°$

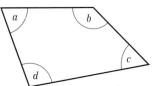

- Opposite angles of a parallelogram are equal.
  $a = c$
  $b = d$

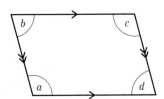

- The **angle of elevation** of point A from point B is the angle of turn above the horizontal to look directly from B to A.

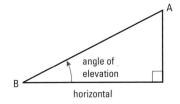

- The **angle of depression** of point B from point A is the angle of turn below the horizontal to look directly from A to B.

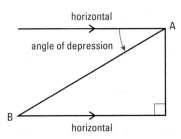

- Angles of elevation and depression are always measured from the horizontal.
- A **polygon** is a closed two-dimensional shape with straight sides.
- A **regular** polygon is a polygon with all its sides the same length and all its angles equal in size. You will need to know the names of some special polygons.
- Sum of the interior angles of a polygon with n sides $= (n - 2) \times 180°$
  $= (2n - 4)$ right angles
- At a vertex of a polygon, interior angle + exterior angle = 180°

- The sum of the exterior angles of any polygon is 360°.
- The radii of a circle are all the same length. This means that a triangle in a circle where two of its sides are radii is an isosceles triangle.

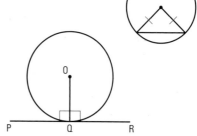

- The angle between a tangent and a radius of a circle is 90°.

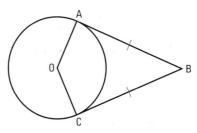

- Tangents to a circle from a point outside the circle are equal in length.

Review exercise

1

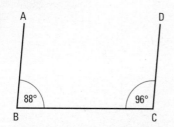

Diagram **NOT** accurately drawn

James says, "The lines AB and DC are parallel."
Ben says, "The lines AB and DC are **not** parallel."
Who is right, James or Ben? Give a reason for your answer.

May 2009

ResultsPlus
Exam Question Report

93% of students answered this question poorly because they did not justify their answers.

2 The diagram shows a building standing on horizontal ground.
  a Work out the size of the angle of elevation of the top of the building from point A on the ground.
  b Write down the size of the angle of depression of point A from the top of the building.

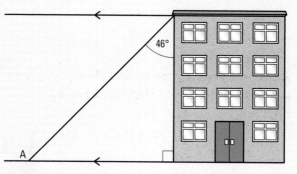

3 Calculate the size of angle *e*. Give a reason for your answer.

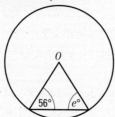

A03

4 ABC is an isosceles triangle.
BCD is a straight line.
AB = AC.
Angle A = 54°.

   a **i** Work out the size of the angle marked x.

      ii Give a reason for your answer.

   b Work out the size of the angle marked y.

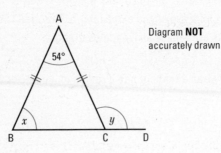

Diagram **NOT** accurately drawn

June 2007

D

5 A bird sits at the top of the tree, looking at a worm at the point P. Write down the angle between the horizontal and the bird's view of the worm.

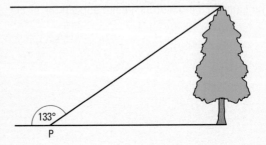

6 T and S are points on the circumference of a circle.
PT and PS are tangents to the circle.
Angle $STP = 52°$.
Angle $TPS = x°$.

   i Work out the value of x.

   ii Give reasons for your answer.

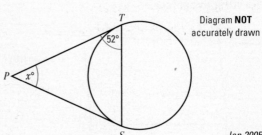

Diagram **NOT** accurately drawn

Jan 2005

7 Find the size of the angle marked x.

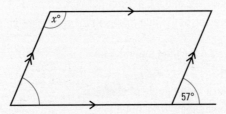

* 8

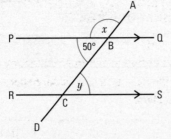

ABCD is a straight line.
PQ is parallel to RS.
Write down the size of angle x and y,
giving reasons for your answer.

March 2008

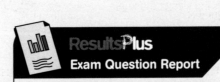

Results **Plus**
Exam Question Report

92% of students answered this question well
because they knew the difference between
alternate and corresponding angles.

9 Work out the value of x.

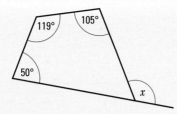

Nov 2007

10 ABCDEF is a regular hexagon and ABQP is a square.
Angle CBQ = $x°$.
Work out the value of x.

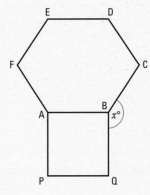

June 2007

11 Work out the size of the angle p.
Give reasons for your working.

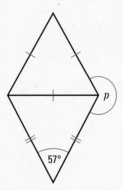

12

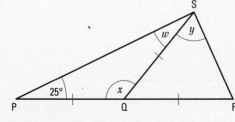

Diagram **NOT**
accurately drawn

PQR is a straight line.
PQ = QS = QR. Angle SPQ = 25°.
a i Write down the size of angle w.
ii Work out the size of angle x.
b Work out the size of angle y.

ResultsPlus
Exam Question Report

84% of students answered this question poorly
because they did not use all of the information
given in the question.

Nov 2008

13 The diagram shows part of a regular 10-sided polygon.
Work out the size of the angle marked x.

Nov 2008

C

14 In the diagram, ABC is a straight line and BD = CD.
  a Work out the size of angle x.
  b Work out the size of angle y.

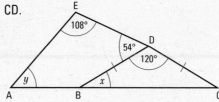

Nov 2006

15 In triangle ABC, AB = AC. Angle ABC = $(x + 20°)$.
Show that angle BAC = $(140° - 2x)$.
Give reasons for each stage of your working.

Diagram **NOT** accurately drawn

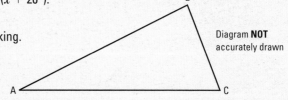

16 In a regular polygon each exterior angle is two thirds the size of each interior angle.
  a Calculate the size of each interior angle.
  b Calculate the number of sides of the polygon.

* 17 In the diagram, AC = BC.
Prove that angle BCD = $4(95° - x)$
Give reasons for each stage of your working.

Diagram **NOT** accurately drawn

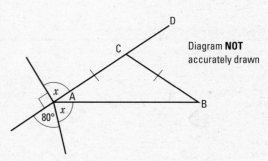

* 18 PQR is a triangle with PQ = PR.
Prove that PY = QY.

Diagram **NOT** accurately drawn

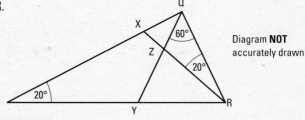

A
A02
A03

19 Just after 1 o'clock the hour and minute hands of a clock are pointing in the same direction, meaning that the angle between them is 0°. What time is this? (Answer to the nearest second).

A*
A02
A03

* 20

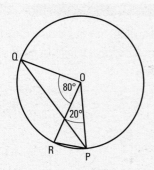

Diagram **NOT** accurately drawn

P, R and Q are points on the circumference of a circle, centre O.
Angle POR = 20°. Angle ROQ = 80°.
Prove that QP bisects angle OPR.

14 AREA AND VOLUME

Compared to humans, elephants have a small surface area in comparison to their mass. This can make it difficult for them to lose heat. One way that they have evolved to solve this problem is by developing large, flapping ears. Elephant ears are made up of a very thin layer of skin stretched over cartilage and a rich network of blood vessels. The large surface area of the ears allows any breeze in the surroundings to flow over these blood vessels and cool the blood down by as much as 10 degrees before returning to the body.

◉ Objectives

In this chapter you will:
- solve problems involving perimeters and areas
- draw the nets, elevations and plans of 3D shapes
- work out the volume and surface area of cuboids and prisms
- specify and find points in three dimensions.

◈ Before you start

You need to know:
- how to measure or calculate the perimeters of rectangles and triangles
- how to use the formula for the area of a rectangle.

14.1 Area of triangles, parallelograms and trapeziums

◎ Objectives

- You know and can use the formula for the area of a triangle.
- You know and can use the formula for the area of a parallelogram.
- You know and can use the formula for the area of a trapezium.

◈ Why do this?

Zoologists at game reserves need to know the areas of different sections of their reserve, so that they know how many animals it can accommodate.

⬆ Get Ready

1. The diagram shows a rectangle.
The length of the rectangle is 9 cm.
The perimeter of the rectangle is 28 cm.
Work out the width and the area of
the rectangle.

9 cm

9 cm

🌐 Key Point

- The **area** of a 2D shape is a measure of the amount of space inside the shape.

Area of a triangle

🌐 Key Points

- The diagram below shows triangle ABC. A rectangle has been drawn around the triangle. The inside of the rectangle has been split into four triangles.

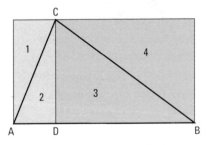

Triangles 1 and 2 are congruent so area triangle 1 = area triangle 2.

Also area triangle 3 = area triangle 4.

- The length of the rectangle is the **base** of the triangle, and the width of the rectangle is the **perpendicular height** of the triangle.

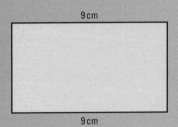
height h

base b

This means that the area of triangle ABC is half the area of the rectangle.

Area of the rectangle = base × height
So to find the area of a triangle, work out a half of its base × its height.

◉ Area of a triangle $= \frac{1}{2} \times$ base $\times$ height

$A = \frac{1}{2} bh$

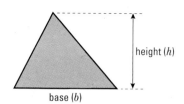

height (h)

base (b)

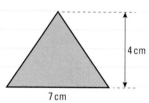 **Example 1**   Work out the area of the triangle.

4 cm

7 cm

Area of a triangle $= \frac{1}{2} \times$ base $\times$ height

$7 \times 4 = 28 \quad \frac{1}{2} \times 28 = 14$

Do not forget to put the units of the answer.

ResultsPlus
Examiner's Tip

The height of a triangle is its vertical or perpendicular height.

Area $= \frac{1}{2} \times 7 \times 4 \, \text{cm}^2$

$= 14 \, \text{cm}^2$

Area of a parallelogram

Key Points

Here are two congruent triangles.

The triangles can be put together to form a parallelogram. The two triangles have equal areas so the area of the parallelogram is twice the area of one of the triangles.

Area of one triangle $= \frac{1}{2} \times$ base $\times$ height

Area of parallelogram $= 2 \times \frac{1}{2} \times$ base $\times$ height $=$ base $\times$ height

◉ Area of a parallelogram $=$ base $\times$ height

$A = bh$

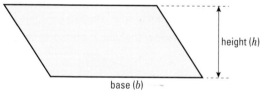

height (h)

base (b)

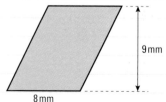 **Example 2**   Work out the area of the parallelogram.

Area $= 8 \times 9 \, \text{mm}^2$

$= 72 \, \text{mm}^2$

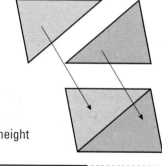

9 mm

8 mm

Area of a parallelogram $=$ base $\times$ height.

As the lengths are in millimetres, the units of the area are mm^2.

Exercise 14A

Questions in this chapter are targeted at the grades indicated.

D

1 Work out the areas of these triangles and parallelograms.

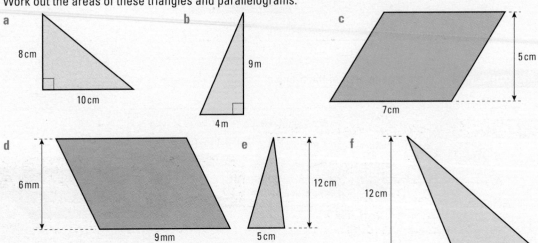

a 8 cm 10 cm

b 9 m 4 m

c 5 cm 7 cm

d 6 mm 9 mm

e 12 cm 5 cm

f 12 cm 9 cm

2 Copy and complete this table.

Shape	Base	Height	Area
Triangle	6 cm	5 cm	
Triangle	5 cm	10 cm	
Triangle		8 cm	24 cm²
Parallelogram	8 cm	4 cm	
Parallelogram	7 cm		56 cm²

Area of a trapezium

Key Points

Here is a trapezium. The trapezium is split into two triangles by a diagonal.
Area of trapezium = area of yellow triangle + area of pink triangle.

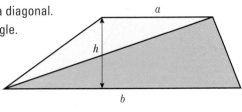

Area = $\frac{1}{2}bh$ h   b

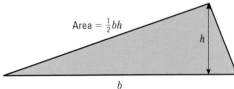

Area = $\frac{1}{2}ah$ h   a

Area of trapezium =
$\frac{1}{2}ah + \frac{1}{2}bh = \frac{1}{2}(a+b)h$

⊚ Area of a trapezium $= \frac{1}{2} \times$ sum of parallel sides $\times$ distance between them.

$A = \frac{1}{2}(a + b)h$

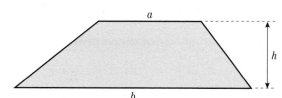

Example 3 Work out the area of the trapezium.

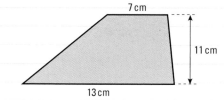

ResultsPlus
Examiner's Tip

Remember that unless the question tells you to take measurements from a diagram you should not do so as diagrams are not accurately drawn.

Area $= \frac{1}{2} \times (7 + 13) \times 11$

⟵ Area of a trapezium =
$\frac{1}{2} \times$ sum of parallel sides $\times$ distance between them.

$= \frac{1}{2} \times 20 \times 11 = 10 \times 11$

Work out the brackets first.

$= 110 \, cm^2$

⚙ **Exercise 14B**

1 Work out the area of each of these trapeziums.

a

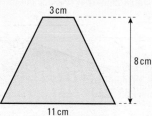

3 cm

8 cm

11 cm

b

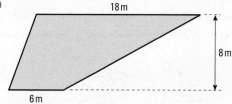

18 m

8 m

6 m

c

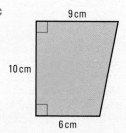

9 cm

10 cm

6 cm

d

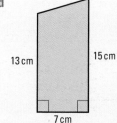

13 cm

15 cm

7 cm

C

14.2 Problems involving perimeter and area

◉ Objectives

● You can find the area and perimeter of a more complicated shape made from simpler shapes.
● You can solve problems involving perimeters and areas.

? Why do this?

A lot of houses seen from the side are a pentagon shape, so a painter would need to work out the area of a pentagon to get the right amount of paint.

◈ Get Ready

1. Write down the formula for the area of:

    a a rectangle   **b** a square   **c** a triangle   **d** a parallelogram   **e** a trapezium.

🔍 Key Point

● The **perimeter** or area of a compound shape can be found by splitting the shape into its simpler parts.

Example 4   Work out the area of this pentagon.

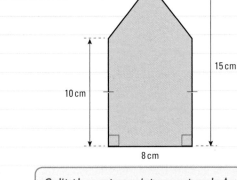

Split the pentagon into a rectangle A and a triangle B.

The height of the triangle is 15 − 10 = 5 cm.
The base of the triangle is 8 cm.

The rectangle has length 8 cm and width 10 cm.

Area of rectangle A = 8 × 10   = 80 cm²
Area of triangle B   = $\frac{1}{2}$ × 8 × 5 = 20 cm²
Area of pentagon   = 80 + 20 = 100 cm²

Area of a rectangle = length × width
Area of a triangle = $\frac{1}{2}$ × base × height
Area of pentagon = area of A + area of B

Example 5

A rectangular wall is 450 cm long and 300 cm high. The wall is to be tiled.
The tiles are squares of side 50 cm. How many tiles are needed?

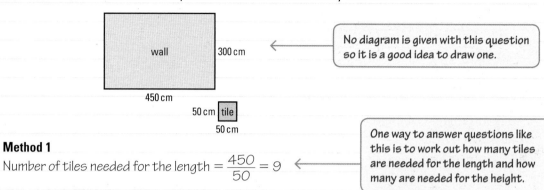

> No diagram is given with this question so it is a good idea to draw one.

Method 1

Number of tiles needed for the length $= \dfrac{450}{50} = 9$

> One way to answer questions like this is to work out how many tiles are needed for the length and how many are needed for the height.

Number of tiles needed for the height $= \dfrac{300}{50} = 6$

> So there are 6 rows of tiles, each with 9 tiles.

Number of tiles needed $= 9 \times 6$

$= 54$

> Number of tiles = number of tiles in each row × number of rows.

Method 2

Area of wall $= 450 \times 300 \text{ cm}^2 = 135\,000 \text{ cm}^2$

Area of a tile $= 50 \times 50 = 2500 \text{ cm}^2$

> The other way to answer this question is to divide the area of the wall by the area of a tile.

Number of tiles $= \dfrac{135\,000}{2500} = 54$

> But remember that you should not use a calculator and the arithmetic is easier in the first method.

Exercise 14C

1 The diagram shows the floor plan of a room.
　a Work out the perimeter of the floor.
　　Give the units of your answer.
　b Work out the area of the floor.
　　Give the units of your answer.

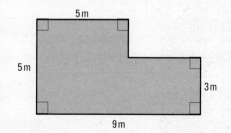

2 Karl wants to make a rectangular lawn in his garden. He wants the lawn to be 30 m by 10 m.
Karl buys rectangular strips of turf 5 m long and 1 m wide.
Work out how many strips of turf Karl needs to buy.

3 A wall is a 300 cm by 250 cm rectangle. The wall is to be tiled.
The tiles are squares of side 50 cm. Work out how many tiles are needed.

4 A rectangle is 9 cm by 4 cm. A square has the same area as the rectangle.
Work out the length of side of the square.

D

5 Keith is going to wallpaper his living room and his bedroom.

Here are the floor plans of these rooms.

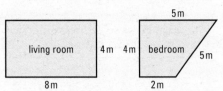

a Work out the area of the floor in:

  i Keith's living room

  ii Keith's bedroom.

b Work out the perimeter of the floor in Keith's living room.

To work out the number of rolls of wallpaper he needs, Keith uses this chart.

Keith is going to use standard rolls of wallpaper.

Standard rolls of wallpaper are approx 10 m long								
How many rolls for the walls								
	Distance around the room including doors & windows							
Wall height	10 m – 33 ft	12 m – 39 ft	14 m – 46 ft	16 m – 52 ft	18 m – 59 ft	20 m – 66 ft	22 m – 72 ft	24 m – 79 ft
2 – 2.3 m 7' – 7'6"	5	5	6	7	8	9	10	11
2.3 – 2.4 m 7'6" – 8'	5	6	7	8	9	10	10	11
2.4 – 2.6 m 8' – 8'6"	5	6	7	9	10	11	12	13
2.6 – 2.7 m 8'6" – 9'	5	6	7	9	10	11	12	13
2.7 – 2.9 m 9' – 9'6"	6	7	8	9	10	12	12	14

A02
A03

The height of the walls in Keith's living room is 2.5 m.

c Find how many rolls of wallpaper Keith needs for his living room.

A02
A03

The height of the walls in Keith's bedroom is 2.6 m.

d Find the number of rolls of wallpaper Keith needs for his bedroom.

C

A02

6 Here is a quadrilateral.

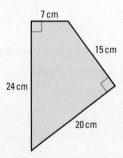

a Work out the perimeter of the quadrilateral.

b Work out the area of the quadrilateral.

7 Work out the area of the yellow shaded region in this diagram.

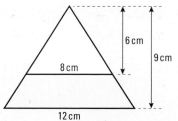
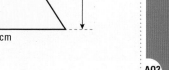

8 A kite has diagonals of length 10 cm and 20 cm.
   Work out the area of the kite.

14.3 Drawing 3D shapes

⊙ Objective

○ You can recognise and draw the net of a 3D shape.

⟷ Why do this?

A manufacturer of chocolate boxes would have to consider the nets of different sizes of boxes in order to see how best to package their product.

⟷ Get Ready

1. Sketch these shapes.
   a a triangular prism   b a square-based pyramid
   c a cylinder   d a triangular-based pyramid

🔍 Key Points

◉ Isometric paper will help you to make scale drawings of **three-dimensional** objects.
◉ Isometric paper must be the right way up i.e. vertical lines down the page and no horizontal lines.
◉ A **net** of a 3D shape is a 2D shape that can be folded to make the 3D shape.
◉ A 3D shape can have more than one net.

This cube has sides of length 2.

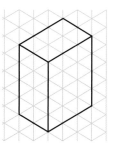

This **cuboid** has height 4, length 3 and width 2.

This prism has a triangular face.

Shapes can be joined together

Example 6 ▶ Draw two different nets for this cuboid.

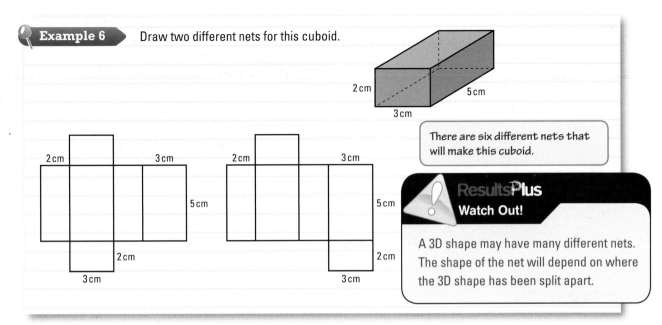

There are six different nets that will make this cuboid.

ResultsPlus
Watch Out!

A 3D shape may have many different nets. The shape of the net will depend on where the 3D shape has been split apart.

Exercise 14D

1 Use isometric paper to draw a cuboid with height 2 cm, width 4 cm and length 3 cm.

2 Sketch six different nets that will make a cube.

3 Here are the nets of some 3D shapes. Identify the shapes.

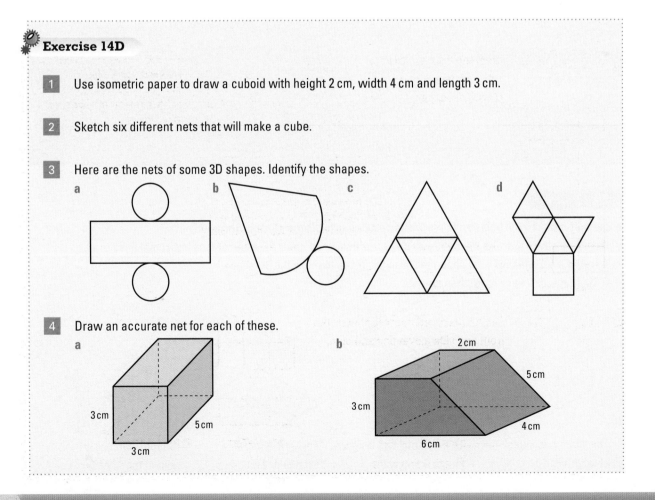

4 Draw an accurate net for each of these.

14.4 Elevations and plans

◎ Objective

○ You can draw elevations and plans of 3D shapes.

◈ Why do this?

Architectural proposals will usually contain plans and elevations of the proposed building, to give people an idea of what the building will look like from each side.

◈ Get Ready

1. What would the shapes in question 4, on page 186, look like if drawn from above, the side and the front?

◈ Key Points

● The **front elevation** is the view from the front.
● The **side elevation** is the view from the side.
● The **plan** is the view from above.

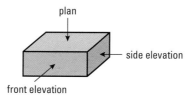

◈ Example 7

Draw the front elevation, side elevation and plan of this 3D shape.

There are six cubes in this shape but you can see only five of them. There must be a cube under the top one.

plan

Draw the elevations and plan like this:
1. Plan at the top.
2. Front elevation under the plan.
3. Side elevation (view from the right) to the right of the front elevation.

front elevation   side elevation

◈ Example 8

Sketch the shape represented by the front and side elevations and plan.

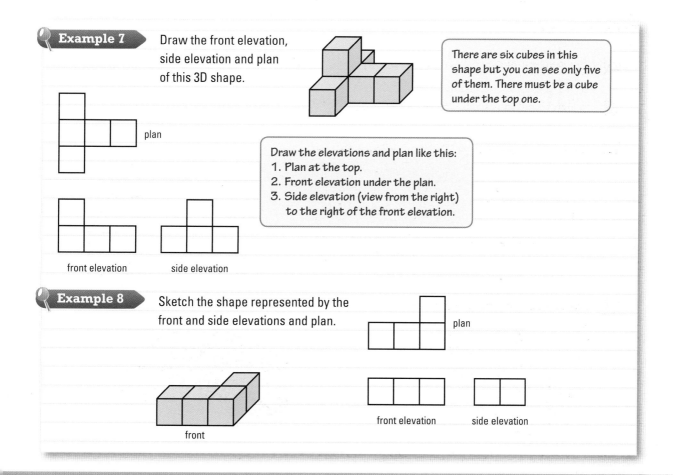

plan

front   front elevation   side elevation

D

Exercise 14E

1 Draw the elevations and plans of these shapes.

a

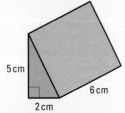

front

b

c

2 m

3 m

5 m

d

5 cm

6 cm

2 cm

e

4 cm

5 cm

3 cm

3 cm

6 cm

f

2 cm

5 cm

g

5 cm

4 cm

2 Sketch the shapes represented by these elevations and plans.

a

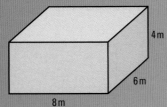

plan

front elevation   side elevation

b

plan

front elevation   side elevation

c

plan

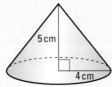

front elevation   side elevation

14.5 Volume of a cuboid

⊙ Objective

○ You can work out the volume of a cuboid and shapes made from cuboids.

⑦ Why do this?

If you were filling a swimming pool you might first have to consider its volume in order to work out how much water you would need.

⬆ Get Ready

1. Work out the volumes of these cuboids. Give the units with your answers.

a

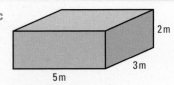

4 m

6 m

8 m

b

6 cm

8 cm

12 cm

Example 9

This shape is made from two cuboids.
Work out the total **volume** of the shape.

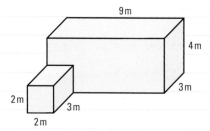

Work out the volume of each cuboid.
Use volume of cuboid $= l \times w \times h$.

For the larger cuboid
$l = 9$ m, $w = 3$ m
and $h = 4$ m.

Volume $= 9 \times 3 \times 4 = 108\,\text{m}^3$

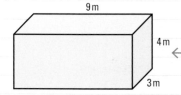

For the smaller cuboid
$l = 2$ m, $w = 3$ m and
$h = 2$ m.

Volume $= 2 \times 3 \times 2 = 12\,\text{m}^3$

Total volume $= 108 + 12 = 120\,\text{m}^3$

To work out the total volume of the shape
add the volumes of the cuboids.

Exercise 14F

1 These shapes are made from cuboids. Work out the volumes of the shapes.

a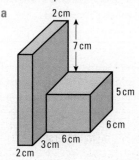

b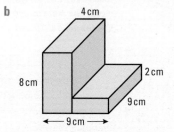

c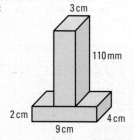

2 Here is a net of a cuboid. Work out the volume of the cuboid.

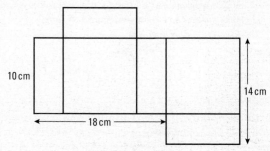

D

A03

14.6 Volume of a prism

 Objective

○ You can work out the volume of a prism.

Why do this?

Sandwiches are often sold in packs that are triangular prisms, so you can work out how much sandwich you are getting.

Get Ready

1. Work out the volume of these shapes.

a

b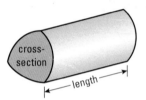

c Find the volume of half shape **b**.

Key Point

● Volume of **prism** = area of cross-section × length

cross-section

length

Example 10 The area of the cross-section of this prism is 25 cm². The length of the prism is 10 cm. Work out the volume of the prism.

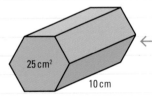

25 cm²

10 cm

Use volume of prism = area of cross-section × length. Here, the area of cross-section = 25 cm² and the length = 10 cm.

Volume = 25 × 10 = 250 cm³

Give the unit with your answer. The unit of area is cm², the length is in cm so the unit of volume is cm³.

Example 11 Work out the volume of this prism.

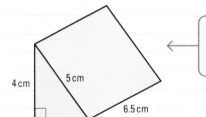

4 cm

5 cm

6.5 cm

3 cm

The cross-section of the prism is a triangle. Remember: area of a triangle = ½ × base × height. Here the base = 3 cm and height = 4 cm.

Area of cross-section = $\frac{1}{2}$ × 3 × 4 = 6 cm²
Volume of prism = 6 × 6.5 = 39 cm³

Use volume of prism = area of cross-section × length. Here the area of cross-section = 6 cm² and length = 6.5 cm.

Exercise 14G

1 Work out the volumes of these prisms.

a

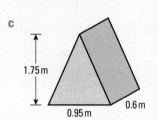

12 cm²   6.5 cm

b

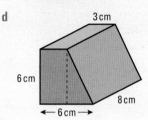

75 mm²   30 mm

c

1.75 m

0.95 m   0.6 m

d

3 cm

6 cm

6 cm   8 cm

2 Work out the volumes of these prisms.

a

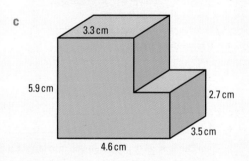

6 cm   5 cm

9 cm   5 cm

b

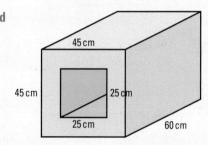

12 cm

15 cm

28 cm   35 cm

c

3.3 cm

5.9 cm   2.7 cm

4.6 cm   3.5 cm

d

45 cm

45 cm   25 cm

25 cm   60 cm

3 The area of the cross-section of a prism is 45 cm². The volume of the prism is 405 cm³.
Work out the length of the prism.

4 Here is a prism. Show that the volume of the prism is $8x^3$ cm³.

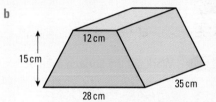

x

$2x$

$2x$

$3x$

5 The diagram shows a triangular prism.
The volume of the prism is $45y^3$ cm³.
Find an expression for h in terms of y.

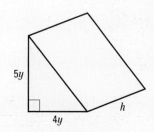

$5y$

$4y$ h

14.7 Surface area of a prism

◎ Objective

• You can work out the surface area of a prism.

② Why do this?

Tents are an example of triangular prisms. Larger tents have a greater surface area of fabric and can accommodate more people.

⬆ Get Ready

1. Work out the area of a triangle with base length 5 cm and height 6 cm.

🔍 Key Point

◉ To work out the **surface area** of a shape, work out the surface area of each of the sides and add them together.

📌 Example 12 Work out the surface area of this cuboid.

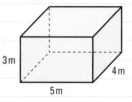

A cuboid has 6 faces. Each face is a rectangle.
Work out the areas of the rectangles.
Opposite faces of a cuboid have the same area so there are 3 repeated areas.

Surface area = $(3 \times 5) + (3 \times 5)$
$\qquad + (3 \times 4) + (3 \times 4) + (5 \times 4) + (5 \times 4)$
$\qquad = 15 + 15 + 12 + 12 + 20 + 20$
$\qquad = 94 \, m^2$

Give the unit with your answer. The lengths of the sides are in m, so the unit of area is m^2.

📌 Example 13 Work out the surface area of this triangular prism.

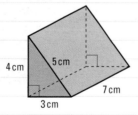

A triangular prism has 5 sides: 2 triangles (with equal areas) and 3 rectangles. Work out the area of the triangular side. Use area of triangle = $\frac{1}{2} \times$ base $\times$ height.
Here base = 3 cm and height = 4 cm.

Surface area = $(\frac{1}{2} \times 3 \times 4) + (\frac{1}{2} \times 3 \times 4)$
$\qquad + (5 \times 7) + (4 \times 7) + (3 \times 7)$
$\qquad = 6 + 6 + 35 + 28 + 21$
$\qquad = 96 \, cm^2$

ResultsPlus
Watch Out!

Make sure you are adding the areas of the right number of sides.

Exercise 14H

1 Work out the surface areas of these shapes. Give the units with your answers.

D

a

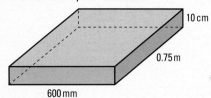
45 cm
55 cm
30 cm

b

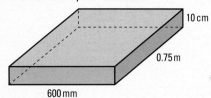
10 cm
0.75 m
600 mm

c

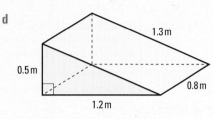
10 cm
14 cm
8 cm
18 cm
35 cm

d

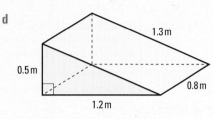
1.3 m
0.5 m
0.8 m
1.2 m

e

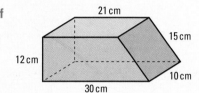
13 cm 12 cm 20 cm
8 cm
21 cm

f

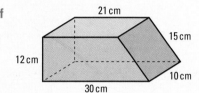
21 cm
15 cm
12 cm
30 cm
10 cm

2 A room has dimensions 6.8 metres × 9.2 metres × 2.5 metres. Stephanie wants to paint the walls of the room. A large tin of paint covers 20 m² and costs £11.75. Ignoring windows and doors, and allowing for two coats of paint, how much will it cost Stephanie to paint the room?

FS   A03   C

14.8 Coordinates in three dimensions

◉ Objective

○ You can use axes and coordinates to specify and find points in three dimensions.

⦿ Why do this?

GPS systems make their calculations in three dimensions.

◈ Get Ready

1. AB is a section of a line between A (6, 1) and B (−3, 5). Find the midpoint.

◈ Key Points

◉ To locate a point in two dimensions, two perpendicular axes are used, the x-axis and the y-axis, and two coordinates are given, the x-coordinate and the y-coordinate.

◉ In three dimensions, an extra axis is needed, the z-axis.

◉ The three axes are perpendicular to each other.

◉ The position of a point is given by three coordinates: the x-coordinate, the y-coordinate and the z-coordinate.

◉ The coordinates of a point are written (x, y, z).

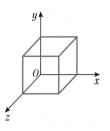

Example 14

The diagram represents a cuboid on a 3D grid.
$OR = 2$ units, $OP = 4$ units and $OS = 3$ units.
Find the coordinates of

  a S **b** P **c** R

  d V **e** U **f** O

  g Find the midpoint of PV.

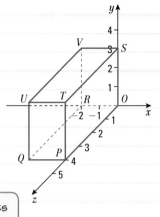

a $S = (0, 3, 0)$ ← To get to S from O you go 0 along the x-axis, 3 units up the y-axis and 0 units parallel to the z-axis.

b $P = (0, 0, 4)$ ← Remember to give the coordinates in the order (x, y, z).

c $R = (-2, 0, 0)$

d $V = (-2, 3, 0)$

e $U = (-2, 3, 4)$ ← To get to U from O you go -2 units along the x-axis, 3 units parallel to the y-axis and 4 units parallel to the z-axis.

f $O = (0, 0, 0)$

g Midpoint of $PV = \left(\dfrac{0 + -2}{2}, \dfrac{0 + 3}{2}, \dfrac{4 + 0}{2} \right) = \left(-1, 1\frac{1}{2}, 2\right)$

Exercise 14I

1   Write down the coordinates of each vertex of this cuboid.

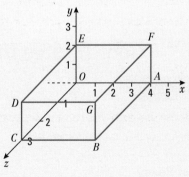

2   Draw a diagram to show the points $A(1, 0, 0)$, $B(1, 0, 3)$ and $C(1, 2, 3)$.

3   **a**   Write down the coordinates of each vertex of this cuboid.

       b   Write down the midpoints of

          i DG

          ii EB

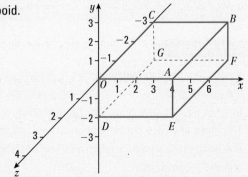

> **4** The coordinates of five of the corners of a cuboid are
> $(1, 0, 0)$, $(1, -3, 0)$, $(1, -3, -1)$, $(1, 0, -1)$ and $(-2, 0, 0)$.
> Find the coordinates of the other three corners.

A03   B

Chapter review

- The area of a 2D shape is a measure of the amount of space inside the shape.
- **Area** of a triangle $= \frac{1}{2} \times$ base $\times$ height.

  $A = \frac{1}{2}bh$
- Area of a parallelogram $=$ base $\times$ height.

  $A = bh$
- Area of a trapezium $= \frac{1}{2} \times$ sum of parallel sides $\times$ distance between them.

  $A = \frac{1}{2}(a + b)h$
- The **perimeter** or area of a compound shape can be found by splitting the shape into its simpler parts.
- Isometric paper will help you to make scale drawings of **three-dimensional** objects.
- Isometric paper must be the right way up i.e. vertical lines down the page and no horizontal lines.
- The **net** of a 3D shape is a 2D shape that can be folded to make the 3D shape.
- A 3D shape can have more than one net.
- The **front elevation** is the view from the front.
- The **side elevation** is the view from the side.
- The **plan** is the view from above.

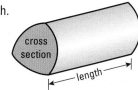

- **Volume** of **prism** $=$ area of cross-section $\times$ length.

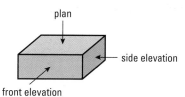

- To work out the **surface area** of a shape, work out the surface areas of each side of the shape and add them together.
- To locate a point in three dimensions, three perpendicular axes are used, the x-axis, the y-axis and the z-axis.
- The position of a point is given by three coordinates: the x-coordinate, the y-coordinate and the z-coordinate.
- The coordinates of a point are written (x, y, z).

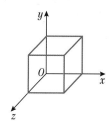

✻ **Review exercise**

1　The diagram shows some nets and some solid shapes.
An arrow has been drawn from one net to its solid shape.
Draw an arrow from each of the other nets to its solid shape.

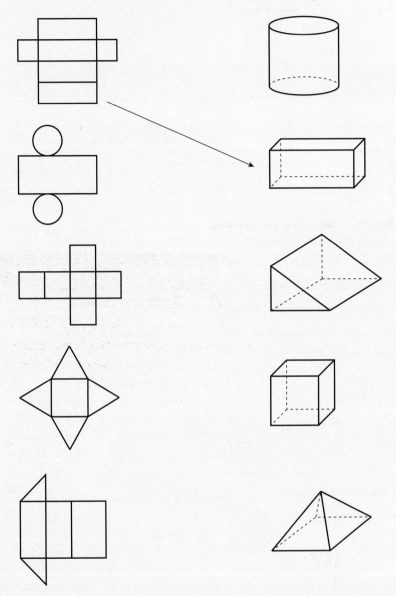

Nov 08

2　Find the volume of this prism.

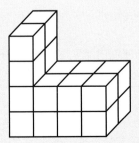

Diagram **NOT**
accurately drawn

represents 1 cm³

June 08

3 On isometric paper, draw the solid represented by these plans and elevations.

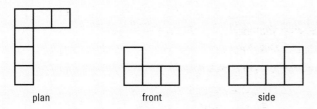

plan   front   side

4 Work out the area of the shape.

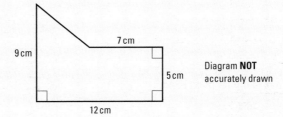

9 cm

7 cm

5 cm

12 cm

Diagram **NOT** accurately drawn

Nov 2008

5 The diagram shows a solid object made of 6 identical cubes.

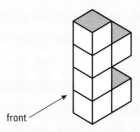

front

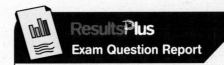

ResultsPlus

Exam Question Report

95% of students answered this question poorly because they did not know what the different types of plans and elevations were.

a On a centimetre grid, draw the side elevation of the solid object from the direction of the arrow.

b On a centimetre grid, draw the plan of the solid object.

June 07

6 The diagram shows a cuboid.

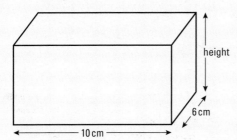

height

6 cm

10 cm

The cuboid has:
a volume of 300 cm³
a length of 10 cm
a width of 6 cm.
Work out the height of the cuboid.

Nov 06

D

7 Boxes are packed into cartons.
A box measures 4 cm by 6 cm by 10 cm.
A carton measures 20 cm by 30 cm by 60 cm.
The carton is completely filled with boxes.
Work out the number of boxes that will
completely fill one carton.

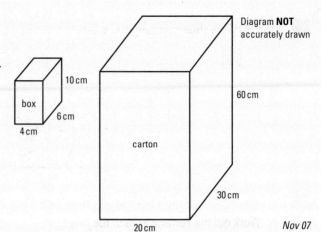

Diagram **NOT**
accurately drawn

10 cm

box

6 cm

4 cm

60 cm

carton

30 cm

20 cm

Nov 07

A02
A03

8 Jane makes chocolates.
Each box she puts them in has:

volume = 1000 cm³
length = 20 cm
width = 1000 cm.

a Work out the height of a box.
Jane makes 350 chocolates.
Each box will hold 18 chocolates.
b Work out: **i** how many boxes Jane can fill completely
ii how many chocolates will be left over.

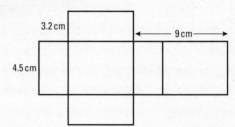

A03

9 Here is a net of a cuboid. Work out:
a the surface area
b the volume of the cuboid.

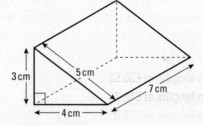

3.2 cm

9 cm

4.5 cm

10 Work out the total surface area of the triangular prism.
Give the units with your answer.

3 cm

5 cm

7 cm

4 cm

June 2008

11 Work out the total surface area of the L-shaped prism.
State the units with your answer.

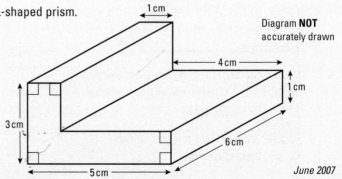

1 cm

Diagram **NOT**
accurately drawn

4 cm

1 cm

3 cm

6 cm

5 cm

June 2007

C

12 The diagram shows a triangular prism.

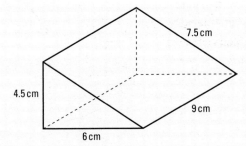

7.5 cm

4.5 cm

9 cm

6 cm

    a Draw the elevations and plan for the prism.
    b Work out the surface area of the prism.
       Give the units with your answer.

13 The diagram shows a farmer's field. The farmer wishes to fence the field up to the corners of the barn, and fertilise the soil.

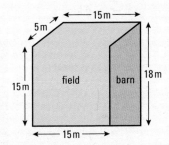

15 m

5 m

15 m

field   barn

18 m

15 m

    a Work out what length of fencing the farmer needs.
    b Work out the area of ground the farmer needs to fertilise.

* **14** Shelim is replacing the skirting boards and coving in his living room.

Skirting board can be bought in:
4 m lengths at £30.50
3 m lengths at £18.75
2 m lengths at £14.00.

Coving can be bought in:
3 m lengths at £27.50
2.4 m lengths at £22.00.

Coving can be joined together, but skirting board must not be pieced together as the joins will be noticeable.

Find the cost of his materials for both jobs, minimising the waste.

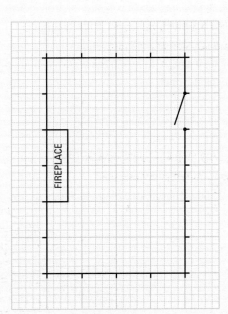

 = 1 m

FS   A02
     A03

FIREPLACE

15 Amy has saved £600 to spend on carpeting her front room. There are four types she likes:

Natural Twist at £14.50 per m²
Medium Blend at £17.60 per m²
Heavy Weave at £19.00 per m²
Luxury Pile at £24.90 per m².

She also needs to buy underlay, which is available in two types:

Cushion at £2.00 per m²
Super Cushion at £4.00 per m².
Fitting is £50 extra.
What can she afford to buy?

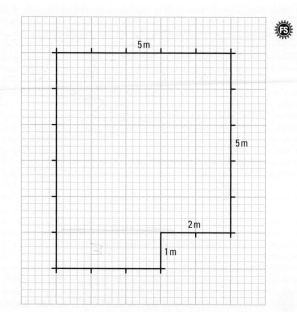

16 A landscape contractor charges:
£40 per square metre for levelling the ground and laying paving stones
£15 per square metre for sowing grass seed.
Calculate the cost of both paving and seeding the garden shown on the right.

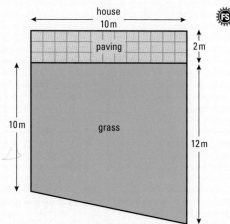

17 A swimming pool has a cross-sectional area in the shape of a trapezium, as shown in the diagram.
Water is pumped in at 2 m³ per minute.
Using the dimensions shown in the diagram, find how long it takes to fill the pool.

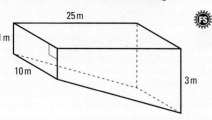

18 The diagram shows a cuboid drawn on a 3D grid.
Vertex A has coordinates (5, 2, 3).
a Write down the coordinates of vertex E.

B and D are vertices of the cuboid.
b Work out the coordinates of the midpoint of BD.

Diagram **NOT** accurately drawn

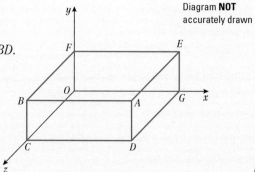

Nov 2008

19 The solid shape, shown in the diagram, is made by cutting a hole all the way through a wooden cube.

The cube has edges of length 7 cm.

The hole has a square cross-section of side 2 cm.

a Work out the volume of wood in the solid shape.

The mass of the solid shape is 189 grams.

b Work out the density of the wood.

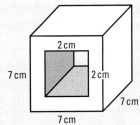

March 2009, adapted

MULTIPLICATION

The following question helps you to develop both your ability to select and apply a method (AO2) and your ability to solve problems using your skills of interpretation (AO3). Your AO3 skills are particularly required as you will need to work through several steps to solve this problem. There are also some functional elements as this is a real-life situation and there is a problem to solve.

Example

Adam runs a coach company. He has 6 small coaches, 4 medium coaches, 3 large coaches and 1 double-decker coach.

The table gives information on how many passengers each coach can seat, the cost of hiring the coach and a driver for a day, and how many of these coaches Adam owns.

Adam's Coach Company			
Coach type	Number of seats	Cost of hire	Number owned
Small	25	£100	6
Medium	38	£110	4
Large	54	£120	3
Double-decker	78	£140	1

Rachel wants to hire some coaches from Adam to take 222 people out for the day. What is the cheapest way for Rachel to do this?

Solution

> As the number of seats increases, the cost goes down proportionally. Therefore you need to use the largest coach, the double-decker, first.

1 double-decker	£140	78
3 large	+ £360	+ 162
	£500	240 seats

> This leaves 144 people to fit in. This could be done with three large coaches but would leave 12 empty seats.

1 double-decker	£140	78
2 large	£240	108
1 medium	+ £110	+ 38
	£490	224 seats

> If two large coaches are used then this would leave 36 people to fit in, so a medium coach would be needed as well

The cheapest way costs £490 with two spare seats.

1 Sam is a salesman. He is paid expenses when he drives his car on company business.

He is paid 45p for each mile he drives.

He is also paid a meal allowance.

Here is Sam's time and mileage sheet for one week.

Meal Allowance
Lunch £8.50
Dinner £22

*Only paid if Sam arrives home after 8 pm

Day	Miles driven	Lunch claimed	Time arrived home
Monday	180	Yes	9 pm
Tuesday	48		5 pm
Wednesday	64	Yes	8.30 pm
Thursday	33		5 pm
Friday	75	Yes	7.30 pm

Work out Sam's total expenses for the week.

2 Lynsey took part in a sponsored swim. Her target was to raise £100 for charity. Her nan promised her that she would make up the £100 if Lynsey did not raise enough.

Here is Lynsey's sponsor form.

Lynsey swam 32 lengths in a pool of length 40 m.

Will her nan have to give her any money?

You must explain your answer.

Sponsor	Amount
Ali	£5
Rob	25p for each length
Will	30p for each length
Mum	50p for each length
Jade	2p for each metre

3 Here are the rates charged for Mr Pitkin's telephone.

Line rental	£29.36
Daytime cost	4p for each minute
Evening and weekend	3p for each minute
To mobiles	11p for each minute
International rate (anytime)	8p for each minute

Here are the details of calls made by Mr Pitkin in one quarter.

Type of call	Minutes
Daytime	78
Evening	312
To mobiles	42
International rate	25

Calculate Mr Pitkin's telephone bill for that quarter.

T he following question helps you develop your ability to select and apply a method (AO2) and your ability to analyse and interpret problems (AO3).

Example   The side of a shed is the shape of a trapezium as shown in the diagram.

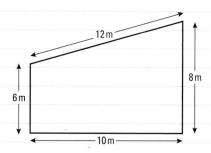

The side is to be given two coats of paint. The paint is sold in 1 litre cans costing £3 each.
1 litre of paint covers 15 square metres.
How much will it cost to paint the wall?

Solution

Using the formula
Area (trapezium) $= \frac{1}{2}(a + b)h$

$$= \frac{1}{2}(6 + 8)10$$

$$= 70\,\text{m}^2$$

> You need to find the area of the side of the shed.
> You may choose to use the formula, or divide the shape into a rectangle and a triangle.
> The formula is given on the formulae sheet.

Dividing the shape
Area of rectangle $= 6 \times 10 = 60\,\text{m}^2$

Area of triangle $= \frac{1}{2} \times 10 \times 2 = 10\,\text{m}^2$

Total area $= 70\,\text{m}^2$

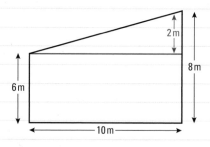

The number of tins of paint required for one coat is $70 \div 15$.

> Total area ÷ area covered by 1 tin

The number of tins needed for two coats is
$140 \div 15 = 9.3$.
So 10 tins will be needed.
The cost of the paint will be £30.

Now try these

1   This shape is made by joining six squares.

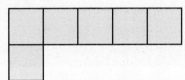

Find two shapes which have the same area but different perimeters.

2   The diagram shows a wall which is to be built with bricks.
The bricks measure 200 mm × 100 mm.
They are sold in packs of 100. One pack costs £35.
Find the cost of the bricks.

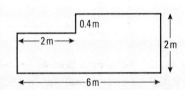

3   The diagram shows a rectangular path around a lawn. The path is 1 m wide.

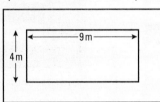

Gravel costs £124 per tonne.
1 tonne of gravel covers 15 m^2.
Work out the cost of covering the path with gravel.

4   Find the perimeter of three different rectangles which each have an area of 36 cm^2.

5   The diagram shows a bathroom wall in the shape of a trapezium. The wall is to be painted.

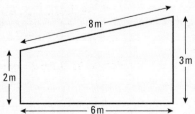

The paint chosen is sold in 1 litre cans costing £4 each. 1 litre covers 12 square metres.
How much will it cost to paint the wall?

COMMUNICATION

In the twenty-first century people communicate using a wide range of technologies. The cost, speed and quality of communication can help determine how and when they are used.

QUESTION

1. When buying a mobile phone the two most popular options are a monthly contract or 'pay as you go'. You estimate that you send about 85 texts and make 5 hours of calls per month. Compare the costs of these two options for a variety of time periods up to 2 years.

MONTHLY CONTRACT

Free phone
100 texts/month
400 min/month
£30/month

PAY-AS-YOU-GO

Phone £90
Texts = 5p each
Calls 7p/min

Using this estimate, compare the monthly costs of these two options. How does this comparison change if you try different estimates for texts sent and calls made?

QUESTION

2. Broadband is the internet access choice for most households but the speed varies greatly. The speed is measured in megabytes per second using the formula:

$$\text{Connection speed (Mb/sec)} = \frac{\text{Data received (Mb)}}{\text{Time (sec)}}$$

Paul's current broadband provides a top speed of 8 megabytes per second but he is thinking of upgrading to 50 megabytes per second. Paul regularly downloads TV programmes from the internet. These typically consist of 2 gigabytes (1 gigabyte = 1024 megabytes) and he does this on average once per week. Estimate how much time per month he will save with the upgraded broadband.

3. Photos contain thousands of little dots called pixels. These pixels are arranged in rows and columns to create the picture. The resolution of the photo when printed or displayed on a screen is stated as the number of pixels per inch (ppi).

A photograph when taken has a fixed number of pixels and as the image is enlarged the resolution gets poorer.

Ranji has a photo of his trip go-karting with 2106 pixels in each row and 1443 pixels in each column. He has three sizes of photo frame to choose from: 13 cm by 19 cm, 15 cm by 23 cm and 20 cm by 30 cm.

Which size should he print the photo at to give him a resolution of at least 200 ppi and the smallest amount of distortion to the photo?

1 cm = 0.39 inches

1443ppi

2106 ppi

LINKS

⊚ For **Question 1** you need to be able to use decimals in calculations. You learnt about this in **Chapter 3**

⊚ You learnt about using formulae in **Chapter 10**. You will need to be able to put figures into a formula f␣ **Question 2**.

⊚ For **Question 3** you need to be able to convert between metric and imperial units of measure. You learnt how to do this in **Chapter 12**.

Fuel bills are one of the largest expenses for homeowners. People can reduce their fuel bills by making their homes more energy efficient.

QUESTION

1. The diagram shows part of the loft of a bungalow. The floor of the loft measures 8 m by 7 m. Gaps of 370 mm are separated by joists which are 30 mm across. Estimate the best cost for insulating the whole loft to a thickness of 100 mm, 150 mm and 200 mm.

8m

7m

370mm ⤒ ⤓ 30mm

Here are some prices at a DIY store.

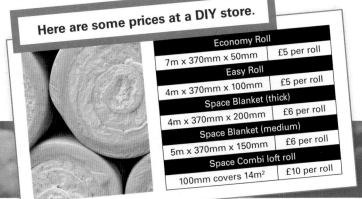

Economy Roll	
7m x 370mm x 50mm	£5 per roll
Easy Roll	
4m x 370mm x 100mm	£5 per roll
Space Blanket (thick)	
4m x 370mm x 200mm	£6 per roll
Space Blanket (medium)	
5m x 370mm x 150mm	£6 per roll
Space Combi loft roll	
100mm covers 14m²	£10 per roll

QUESTION

2. Energy-saving lightbulbs can save significant amounts of money.
They are best used where lights are left on for more than an hour at a time.
Mary decides to replace the old bulbs in her hall light, which has two 60 watt bulbs, and her living room light, which has three 100 watt bulbs, with energy efficient bulbs.

20 watt energy efficient bulbs are the equivalent of 100 watt ordinary bulbs.
11 watt energy efficient bulbs are the equivalent of 60 watt bulbs. The cost of electricity is worked out from the number of units, E, used. The formula for E is:

$$E = \frac{pt}{1000} \text{ where } p \text{ is the power in watts and } t \text{ is the time in hours}$$

In the winter, her hall light is usually on from 5pm until 11pm and her living room light is on for an average of 5 hours a night. Her electricity provider bills her every 13 weeks for a quarterly period, and charges her 11.47p per unit. How much money can she expect to save on her winter bill?

All new appliances come with energy labels that provide you with information on the efficiency of the product.

3. Isaac uses his washing machine five times a week. He is currently being charged 14p per unit for his electricity. The energy consumption per cycle is the number of units used when one complete cycle is done. Isaac is considering replacing his current machine with the new Spinner Max. How long would it take for the running cost savings to exceed the initial purchase price?

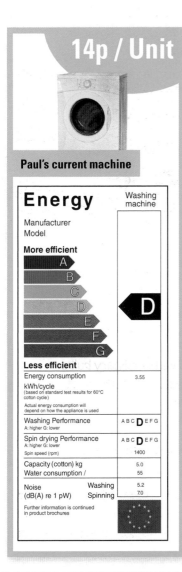

Paul's current machine

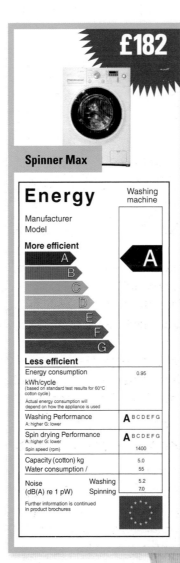

Spinner Max

LINKS

For **Question 1** you will need to convert between different units of measure. You learnt how to do this in **Chapter 12**.

You learnt how to use formulae in **Chapter 10**. You will need to be able to do this in **Question 2**.

For **Question 3** you will need to be able to use decimals in your calculation. You learnt how to do this in **Chapter 3**.

Answers

Chapter 1 Answers

1.1 Get Ready

1 a Neither **b** Factor **c** Factor **d** Multiple
e Neither **f** Factor
2 a No **b** Yes **c** No **d** Yes **e** No

Exercise 1A

1 Yes, for example $2 + 3 = 5$ is prime.
2 4
3 $n = 2, m = 3, p = 7$
4 a 2, 24 **b** 5, 10 **c** 2, 20 **d** 6, 18
5 a $24 = 2^3 \times 3, 60 = 2^2 \times 3 \times 5$
b 12 **c** 120
6 a $72 = 2^3 \times 3^2, 120 = 2^3 \times 3 \times 5$
b 24 **c** 360
7 a 18, 180 **b** 18, 216 **c** 12, 480 **d** 36, 720
8 a 6 **b** 2520
9 a 40 **b** 126 000
10 a 11, 13, 17 and 19 are prime numbers between 10 and 20.
b 23, 29, 31 and 37 are prime numbers between 20 and 40.
c 37, 41, 43, 47, 53, 59, 61 and 67 are prime numbers between 34 and 68.
11 Every 2 minutes
12 3 boxes of burgers and 4 packets of buns
13 No. If one of the prime numbers is 2 you will get an even number.

1.2 Get Ready

1 a 36 **b** 8 **c** 9

Exercise 1B

1 a 1, 4, 9, 16, 25, 36, 49, 64, 81, 100, 121, 144, 169, 196, 225
b 1, 8, 27, 64, 125
2 a i 64, 1, 49, 9 **ii** 64, 1, 8
b i 4, 16 **ii** 125, 27
c i 64, 81, 144 **ii** 125, 64
d i 100, 81, 169, 64 **ii** 125, 64

Exercise 1C

1 a 9 **b** 49 **c** 64
d 1000 **e** 121
2 a 6 **b** 4 **c** 9
d 1 **e** 8
3 a 36 **b** -8 **c** 81
d -1 **e** 144
4 a 2 **b** -3 **c** -1
d 4 **e** 10
5 a 17 **b** 50 **c** 250
d 14 **e** 1 **f** 8
g 34 **h** -11 **i** 9
j 50 **k** 10 **l** 6

1.3 Get Ready

1 a 18 **b** 16 **c** 10

Exercise 1D

1 a 25 **b** 13 **c** 6 **d** 4
e -5 **f** 8 **g** 6 **h** 2
i 1 **j** 5 **k** 14 **l** 52
m 32 **n** 15 **o** 8 **p** 70
2 a 49 **b** 25 **c** 243 **d** 123
e 72 **f** 69 **g** 7 **h** -7
3 a 25 **b** 4 **c** 8 **d** 31

1.4 Get Ready

1 32
2 125
3 729

Exercise 1E

1 a 6^{12} **b** 4^5 **c** 7^6
d 5^6 **e** 3^{10}
2 a 100 000 **b** 125 **c** 64
d 9 **e** 64
3 a 5 **b** 3 **c** 5
d 4 **e** 9
4 a 3^4 **b** 5^9 **c** 2^6
d 6^5 **e** 4^2
5 a 9 **b** 16 **c** 16
d 10 000 **e** 49
6 a 3 **b** 5 **c** 2
d 4 **e** 3

Review exercise

1 a 96 **b** 33 **c** 23
2 a -2 **b** 2 **c** -4
3 a 100 cans **b** 14 (with 4 cans spare)
c 200 cans **d** 14 (with 8 cans spare)
4 a 5 arrangements: 1 by 36, 2 by 18, 3 by 12, 4 by 9, 6 by 6
b 3 arrangements: 1 by 18, 2 by 9, 3 by 6
c 3 arrangements: 1 by 12, 2 by 6, 3 by 4
5 64 is the next number which is both a square number and a cube number.
6 a 5 760 000 **b** 5 760 000 **c** 160
d 57 600 **e** 2304
7 a 3 **b** 30 **c** 0
8 a 3^2 **b** 4 **c** 2^{12} **d** 5^2
9 a $2^2 \times 3^2 \times 7$
b $2^3 \times 3^3 \times 7$
10 9, 15
11 $84 = 2^2 \times 3 \times 7$
$168^2 = 2^6 \times 3^2 \times 7^2$
12 120 000 miles
13 a 48 **b** 36 **c** 343 **d** 25
14 3

15 a False, e.g. $3 + 5 = 8$
  b False, e.g. $4 + 9 = 13$
  c False, e.g. $5 - 3 = 2$
  d False, e.g. $2 \times 3 = 6$
  e True
16 a $2^7 = 128$ **b** 2020

Chapter 2 Answers

2.1 Get Ready

1 $\frac{8}{9}$ **2** $\frac{19}{8}$ **3** $9\frac{2}{5}$

Exercise 2A

1 a $\frac{8}{11}$ **b** $\frac{5}{9}$ **c** $\frac{11}{15}$ **d** $\frac{2}{5}$
2 a $\frac{7}{10}$ **b** $\frac{10}{21}$ **c** $\frac{29}{35}$ **d** $\frac{8}{9}$
  e $\frac{13}{20}$ **f** $\frac{13}{18}$ **g** $\frac{7}{18}$ **h** $\frac{3}{4}$
3 a $\frac{1}{4}$ **b** $\frac{1}{12}$ **c** $\frac{19}{40}$ **d** $\frac{2}{9}$
  e $\frac{1}{4}$ **f** $\frac{5}{12}$ **g** $\frac{1}{10}$ **h** $\frac{1}{18}$
4 a $1\frac{11}{40}$ **b** $1\frac{11}{20}$ **c** $\frac{1}{6}$ **d** $\frac{19}{20}$
  e $1\frac{11}{15}$ **f** $1\frac{5}{8}$ **g** $1\frac{1}{5}$ **h** $1\frac{5}{18}$

Exercise 2B

1 a $8\frac{1}{4}$ **b** $5\frac{3}{10}$ **c** $11\frac{5}{42}$ **d** $18\frac{3}{20}$
2 $7\frac{1}{12}$ miles **3** $1\frac{22}{35}$ lb

Exercise 2C

1 a $1\frac{1}{4}$ **b** $2\frac{3}{8}$ **c** $\frac{3}{4}$ **d** $3\frac{1}{3}$
2 a $\frac{3}{4}$ **b** $1\frac{7}{12}$ **c** $2\frac{24}{35}$ **d** $3\frac{4}{9}$
3 $3\frac{3}{8}$ kg
4 $2\frac{7}{8}$ pints

2.2 Get Ready

1 32 **2** 45 **3** $\frac{19}{5}$ **4** $4\frac{2}{3}$

Exercise 2D

1 a $\frac{3}{10}$ **b** $\frac{3}{20}$ **c** $\frac{6}{11}$ **d** $\frac{2}{9}$
  e $\frac{4}{21}$ **f** $\frac{9}{20}$ **g** $\frac{3}{10}$ **h** $\frac{15}{32}$
2 a $\frac{2}{3}$ **b** $\frac{3}{4}$ **c** $3\frac{3}{5}$ **d** 15
3 a 21 kg **b** $6\frac{2}{3}$ m **c** $7\frac{1}{2}$ litres **d** $7\frac{1}{2}$ pints
4 21
5 £32.65
6 a $\frac{5}{12}$ **b** $\frac{4}{5}$ **c** $4\frac{1}{8}$ **d** 7
  e 3 **f** $4\frac{3}{8}$ **g** 10 **h** $22\frac{1}{2}$
7 $14\frac{5}{8}$ minutes
8 $20\frac{5}{8}$ lb

2.3 Get Ready

1 $\frac{6}{35}$ **2** $\frac{1}{6}$ **3** $\frac{24}{7}$

Exercise 2E

1 a $\frac{5}{12}$ **b** $\frac{3}{16}$ **c** $2\frac{2}{3}$ **d** $1\frac{1}{2}$
  e $\frac{3}{4}$ **f** $1\frac{1}{5}$ **g** $1\frac{11}{14}$ **h** $\frac{5}{6}$
2 a $\frac{1}{2}$ **b** 28 **c** $2\frac{1}{12}$ **d** $2\frac{4}{13}$
  e 2 **f** $3\frac{2}{3}$ **g** 6 **h** $1\frac{7}{8}$
3 16
4 36
5 $11\frac{2}{11}$ days

2.4 Get Ready

1 $\frac{7}{8}$ **2** $\frac{4}{45}$ **3** $3\frac{5}{8}$

Exercise 2F

1 a $\frac{5}{6}$ **b** $\frac{1}{6}$
2 880 m³
3 £5.60
4 £78
5 525
6 $\frac{3}{8} \times 36 = 13\frac{1}{2}$ which is not a whole number
7 $4\frac{1}{12}$ hours
8 $\frac{1}{2}$
9 $\frac{7}{8}$ km
10 192
11 80

Review exercise

1

Name	Hourly rate	Overtime at time and a half	Overtime at double time
Aaron	£8.50	£12.75	£17.00
Chi	£12.00	£18.00	£24.00
Mahmood	£14.40	£21.60	£28.80

2 a $\frac{13}{20}$ **b** $1\frac{7}{12}$ **c** $5\frac{7}{12}$ **d** $84\frac{13}{24}$
3 a £374 **b** 6 hours
4 a $2\frac{4}{7}$ **b** $1\frac{1}{9}$ **c** $1\frac{1}{2}$ **d** $\frac{7}{15}$
5 a $13\frac{1}{2}$ **b** $9\frac{3}{5}$ **c** $3\frac{13}{15}$ **d** $16\frac{7}{26}$
6 a $\frac{1}{12}$ **b** $3\frac{13}{30}$ **c** $1\frac{11}{12}$
7 a $4\frac{1}{6}$ m² **b** $8\frac{1}{3}$ m **c** $\frac{5}{6}$ m
8 Yes, the part is $6\frac{2}{16}$ cm long.
9 720 000 m²
10 $3\frac{1}{12}$ miles
11 a $8\frac{7}{16}$ inches **b** $11\frac{1}{4}$ inches
12 $a = \frac{8}{15}, b = \frac{1}{3}, c = \frac{1}{5}, d = \frac{4}{15}$
13 64
14 350
15 a $\frac{1}{4}$ **b** $\frac{1}{8}$ **c** $\frac{1}{4}$ **d** $\frac{3}{8}$

Chapter 3 Answers

3.1 Get Ready

1 8.02, 8.09, 8.092, 8.2, 8.29, 8.9, 8.92
2 a $\frac{1}{3}$ **b** $\frac{7}{12}$ **c** $\frac{7}{8}$

Exercise 3A

1 0.8, 0.85, $\frac{86}{100}$, $\frac{9}{10}$, 0.98
2 a terminating   **b** terminating   **c** recurring
   d recurring   **e** terminating   **f** recurring
3 Mitch is correct, as there is a factor of 3 in the denominator.

3.2 Get Ready

1 a 18.79   **b** 5.18   **c** 32.74

Exercise 3B

1 a 0.12   **b** 0.0012   **c** 0.04   **d** 0.0063
2 a 2.536   **b** 1.263   **c** 0.043 38   **d** 2.52
   e 13.02   **f** 0.504   **g** 0.046 72   **h** 0.323
3 15p
4 £6.86
5 a 60   **b** 14   **c** 640   **d** 65
   e 25   **f** 2040   **g** 0.05   **h** 0.092
6 a 2.31   **b** 642   **c** 41.3   **d** 42.2
7 £26.13
8 5

3.3 Get Ready

1 a 0.5772   **b** 160.$\dot{3}$ **c** £1.41

Exercise 3C

1 a 6.4   **b** 5.7   **c** 16.9
   d 0.1   **e** 1.0
2 a 5.67   **b** 8.06   **c** 0.13
   d 3.04   **e** 0.08
3 a 6.446   **b** 0.079   **c** 5.079
   d 6.008   **e** 0.020

3.4 Get Ready

1 a 3   **b** 0   **c** 9
2 e.g. 438, 48, 6798

Exercise 3D

1 a 3900   **b** 230   **c** 46
   d 6.5   **e** 5.1   **f** −0.43
2 a 2500   **b** 39.0   **c** 4.90
   d 4.09   **e** 0.0110
3 a 3000   **b** 40   **c** 3
   d 8   **e** 20   **f** 1

3.5 Get Ready

1 a 5000   **b** 20   **c** −7
2 a 600   **b** 15 000   **c** 540 000
3 a 60   **b** 300   **c** 0.025

Exercise 3E

1 a 4200   **b** 7000   **c** 6000
   d 200 000   **e** 80 000
2 a 35   **b** 2   **c** 10   **d** 5   **e** 6
3 a 8000, overestimate   **b** 4, overestimate
   c 10, underestimate   **d** 300, overestimate
4 a 50, overestimate   **b** 4, underestimate
   c 40, overestimate   **d** 8, overestimate
5 2400

3.6 Get Ready

1 a 0.3   **b** 0.05   **c** 0.001
2 a 6   **b** 150   **c** 0.054
3 a 6000   **b** 30 000   **c** 25

Exercise 3F

1 a 2.4   **b** 0.15   **c** 1   **d** 0.0018
2 a 20   **b** 50   **c** 100   **d** 4
   e 0.2
3 a 1.5, underestimate   **b** 0.2, overestimate
   c 5, underestimate   **d** 200, underestimate
4 a 400, overestimate   **b** 210, overestimate
   c 30, underestimate   **d** 4000, underestimate
5 0.25

3.7 Get Ready

1 a 60   **b** 600   **c** 6000
2 a 30   **b** 3   **c** 0.3

Exercise 3G

1 a 1792   **b** 1792   **c** 17.92   **d** 0.017 92
2 a 146.4   **b** 1.464   **c** 0.1464   **d** 0.014 64
3 a 348   **b** 3480   **c** 34.8   **d** 348
4 a 128.8   **b** 230   **c** 56   **d** 2
5 a 0.026   **b** 340   **c** 0.034   **d** 100
6 a 13   **b** 13   **c** 13 000   **d** 0.065

Review exercise

1 £153.90
2 $\frac{2}{3}$
3 0.47, $\frac{12}{25}$, $\frac{3}{5}$, $\frac{31}{50}$
4 £42.96
5 a 4780   **b** 107   **c** 3.23×10^{15}
   d 7000   **e** 57.0
6 a 46   **b** 31   **c** 0.046
   d 20   **e** 4.1
7 a 400   **b** 40   **c** 1×10^{18}
   d 0.005   **e** −3
8 a 3.1   **b** 0.6   **c** 2.1   **d** 4.0
9 a $\frac{77}{400}$ **b** $\frac{77}{4} = 19\frac{1}{4}$ **c** $\frac{770\,000}{4} = 192\,500$
10 75 923 1p pieces
11 a Students' checks
   b

	A	B	C	Best deal
John	£56.75	£62.27	£64.33	Avery Energy
Vijay	£192.37	£186.17	£280.42	Brawn Power

12 a 3000   b 4000   c 24
    d 350 000   e 360
13 a 5   b 5   c 80
    d 250   e 1
14 36 minutes
15 a $\frac{3}{1}$ = £3   b $\frac{30}{1.5}$ = £20   c $\frac{1500}{150}$ = £10
    d $\frac{10}{2.5}$ = £4   e $\frac{300}{50}$ = £6   f $\frac{2000}{10}$ = £200
16 a £21   b 32
17 a 75, underestimate   b 150, underestimate
    c 3.6, overestimate   d 6000, overestimate
    e 45 000, overestimate
18 Volume ≈ 3 × 8 × 9 = 216 m³.
    Number of people = $\frac{216}{4}$ = 54
19 £370.22

Chapter 4 Answers

4.1 Get Ready

1 a i $\frac{1}{5}$ ii 0.2   b i $\frac{1}{8}$ ii 0.125
   c i $\frac{3}{5}$ ii 0.6   d i $\frac{7}{40}$ ii 0.175

Exercise 4A

1 a £180   b 6 kg   c £1.62   d 4.96 kg
   e £6   f 45   g 2.52 km   h £52.50
2 12
3 £10
4 1058
5 $\frac{7}{22}$
6 £17 800
7 Country B, £500 less

4.2 Get Ready

1 £75   2 £16   3 $\frac{1}{4}$

Exercise 4B

1 a £360   b 117 kg   c 50 km   d £1380
2 £220.50
3 £368 000
4 £752
5 £22 470
6 1296
7 Jamil £1400; Sam £1218 (approx – using 4 weeks as 1 month)
   therefore Jamil earns the biggest salary
8 Sarah £4200; Jack £4230 therefore Jack has the most money

Exercise 4C

1 a £360   b 170 kg   c 49 m   d £825
2 £1700
3 1485
4 £455
5 £714
6 89.3 kg

Review exercise

1 a 360   b 22%
2 £94
3 a £240   b 5 kg   c 10.5 kg   d £10.50

4   £34
5   18.54 litres per day
6   16.7%
7   A (A £510, B £512, C £511.13)
8   CompuSystems (Able £23 000, Beta £23 400,
    CompuSystems £24,240, Digital £24 000)
9   £7800
10 14.2%

Chapter 5 Answers

5.1 Get Ready

1 $\frac{1}{16}$ 2 $\frac{9}{25}$ 3 −8

Exercise 5A

1 a 1   b $\frac{1}{8}$ c $\frac{1}{5}$ d 1
   e $-\frac{1}{8}$ f $\frac{1}{81}$ g $\frac{1}{10\,000}$ h 1
   i $\frac{1}{9}$ j 1   k 1   l $\frac{1}{1\,000\,000}$
2 a 3   b $\frac{7}{2}$ c 49   d 64
   e 16   f $15\frac{5}{8}$ g 1   h $\frac{5}{9}$
   i $\frac{25}{49}$ j $\frac{27}{64}$ k 10 000   l 125

5.2 Get Ready

1 a 1000   b $\frac{-1}{100}$
2 10^4
3 23 500

Exercise 5B

1 a 7×10^5 b 6×10^2 c 2×10^3
   d 9×10^8 e 8×10^4
2 a 600 000   b 10 000   c 800 000
   d 300 000 000 e 70
3 a 4.3×10^4 b 5.61×10^5 c 5.6×10^1
   d 3.47×10^1 e 6×10^1
4 a 39 600   b 68 000 000   c 8020
   d 57   e 9.23
5 7×10^9
6 4×10^4

Exercise 5C

1 a 5×10^{-3} b 4×10^{-2} c 7×10^{-6}
   d 9×10^{-1} e 8×10^{-4}
2 a 0.000 06   b 0.08   c 0.000 000 5
   d 0.3   e 0.000 000 01
3 a 4.7×10^{-3} b 9.87×10^{-1} c 8.034×10^{-4}
   d 1.5×10^{-4} e 6.01×10^{-1}
4 a 0.000 084 3 b 0.0201   c 0.000 000 42
   d 0.078 54   e 0.000 94
5 a 4.57×10^5 b 2.3×10^{-3} c 3×10^{-4}
   d 2.356×10^6 e 7.82×10^{-1} f 8.9×10^4
   g 2×10^2 h 5.26×10^{-3} i 6.034×10^3
   j 8.73×10^6
6 a 0.000 412 b 3000   c 20 650 000
   d 0.000 004 e 327 000 000 f 0.75
   g 156.23   h 0.000 000 512 i 270 000
   j 0.612
7 1×10^{-6} m
8 6.25×10^{-2} mm

Answers

Exercise 5D

1 **a** 4.5×10^4 **b** 9.8×10^{-1}
    c 3.4×10^1 **d** 1.86×10^{12}
2 **a** 9×10^2 **b** 4.5×10^4
    c 3.708×10^{-13} **d** 6×10^{-10}
3 **a** In standard form
    b 8.9×10^8 **c** 1.32×10^{-4} **d** 5.6×10^8
    e 6×10^{-4} **f** In standard form
    g 4.005×10^{-12} **h** 9.08×10^{18}
    i In standard form   **j** 4.6×10^5
    k 6.7×10^1 **l** 4×10^0
4 $6\,290\,000, 6.3 \times 10^6, 63.4 \times 10^5, 0.637 \times 10^7$
5 $0.000\,033, 3.35 \times 10^{-5}, 0.034 \times 10^{-2}, 37 \times 10^{-4}$

Exercise 5E

1 **a** 8×10^{11} **b** 9×10^8 **c** 1.2×10^{-1}
    d 4.3×10^{-5} **e** 2.5×10^8 **f** 1×10^{-4}
2 **a** 4×10^{10} **b** 2.5×10^{-9} **c** 1.6×10^{13}
    d 4.9×10^{-15}
3 $3.2 \times 10^0 \text{ mm}^2$

5.3 Get Ready

1 10 **2** 2 **3** -3

Exercise 5F

1 **a** 3 **b** 7 **c** 10 **d** 2 **e** $\frac{1}{2}$
2 **a** 3 **b** 10 **c** -4 **d** 5 **e** $\frac{1}{2}$
3 **a** $\frac{1}{2}$ **b** $\frac{1}{2}$ **c** $\frac{1}{5}$ **d** 2 **e** $\frac{3}{2}$
4 **a** 9 **b** 100 **c** 16 **d** 8 **e** 125
5 **a** $\frac{1}{25}$ **b** $\frac{1}{1000}$ **c** $\frac{1}{3}$ **d** $\frac{1}{4}$ **e** $\frac{1}{512}$
    f $\frac{1}{625}$ **g** $\frac{2}{25}$
6 **a** $n = -1$ **b** $n = 6$ **c** $n = -\frac{1}{2}$
    d $n = \frac{5}{2}$ **e** $n = \frac{11}{3}$

5.4 Get Ready

1 $1, 4, 9, 16, 25, 36, 49, 64, 81, 100$
2 **a** 6 **b** 10
3 $\sqrt{9}, \sqrt{64}$

Exercise 5G

1 **a** 2 **b** 3 **c** 5 **d** 4
2 **a** $10\sqrt{2}$ **b** $4\sqrt{2}$ **c** $2\sqrt{5}$ **d** $2\sqrt{7}$
3 $x = \pm\sqrt{30}$
4 **a** $3 + 2\sqrt{3}$ **b** $5 + 3\sqrt{3}$ **c** $3 + \sqrt{5}$
    d $-5 + \sqrt{7}$ **e** $7 - 4\sqrt{3}$ **f** $27 + 10\sqrt{2}$
5 $2\sqrt{10} \text{ cm}$
6 **a** 12 cm **b** 4 cm^2
7 $3 + 2\sqrt{2} \text{ cm}^2$

Exercise 5H

1 **a** $\frac{\sqrt{2}}{2}$ **b** $\frac{\sqrt{5}}{5}$ **c** $\frac{\sqrt{10}}{2}$
    d $\sqrt{2}$ **e** $\frac{2\sqrt{3}}{3}$
2 **a** $1 + \sqrt{2}$ **b** $-1 + 3\sqrt{2}$ **c** $1 + 2\sqrt{5}$
    d $-1 + 4\sqrt{3}$ **e** $1 + 2\sqrt{7}$

3 $\frac{3\sqrt{6}}{2} \text{ cm}^2$
4 **a** $x = 3 \pm \sqrt{7}$ **b** $x = -5 \pm \sqrt{11}$
5 **a** 1.5 cm^2 **b** $\sqrt{22} \text{ cm}$

Review exercise

1 **a** 4000 **b** $260\,000\,000$ **c** 370
    d 0.0009 **e** 0.135 **f** 0.00008001
2 **a** 1 **b** $\frac{1}{4}$ **c** 1 **d** $\frac{1}{8}$
3 **a** 1 **b** 1 **c** $\frac{1}{9}$ **d** 1
4 **a** 3 **b** 3 **c** $\frac{1}{2}$ **d** 8
5 **a** 3 **b** 10 **c** 2 **d** 4
6 **a** $\frac{1}{3}$ **b** $\frac{1}{7}$ **c** $\frac{1}{5}$ **d** $\frac{1}{2}$
7 **a** 2 **b** $\frac{1}{2}$
8 **a** **i** 7.9×10^3 **ii** 3.5×10^{-4}
    b 5×10^7
9 9.3×10^8
10 $n = \frac{5}{2}$
11 **a** $k = \frac{3}{2}$ **b** $m = 16$ **c** $\frac{\sqrt{2}}{32}$
12 **a** $x = -2$ **b** $x = -4$ **c** $x = 3$ **d** $x = -\frac{1}{2}$
13 Using $a^2 - b^2 = (a + b)(a - b)$
$$\frac{1}{\sqrt{2} + 1} + \frac{1}{\sqrt{3} + \sqrt{2}} + \frac{1}{\sqrt{4} + \sqrt{3}} + \ldots + \frac{1}{10 + \sqrt{99}}$$
$$= \frac{\sqrt{2} - 1}{2 - 1} + \frac{\sqrt{3} - \sqrt{2}}{3 - 2} + \frac{\sqrt{4} - \sqrt{3}}{4 - 3} + \ldots + \frac{10 - \sqrt{99}}{100 - 99}$$
$$= -1 + 10 = 9$$

Chapter 6 Answers

6.1 Get Ready

1 18 km

Exercise 6A

1 **a** $1:4$ **b** $5:3$ **c** $7:9$ **d** $11:4$
2 **a** $1:3$ **b** $1:4$ **c** $1:3.5$ **d** $1:0.5$
    e $1:0.3$ **f** $1:0.6$ **g** $1:8$ **h** $1:\frac{8}{15}$
3 $1:5$
4 **a** $\frac{4}{9}$ **b** $4:5$ **c** $1:1.25$
5 $1:375$
6 **a** $1:\frac{1}{6}$ **b** $1:0.2$ **c** $1:0.02$ **d** $1:40$

6.2 Get Ready

1 **a** $2:3$ **b** $1:5$ **c** $8:7$ **d** $9:200$

Exercise 6B

1 **a** 10 g **b** 30 g **c** 250 g
2 **a** 10 kg **b** 15 kg
3 108.5 km
4 £1080
5 1.53 m
6 250

6.3 Get Ready

1 20.65 **2** 6 **3** 241.5

Exercise 6C

1. a £4.26 : £10.65   b 360 g : 240 g
   c £14.21 : £56.84 : £99.47   d 6.3 m : 12.6 m : 15.75 m
2. 60°, 50°, 70°
3. £36.50
4. $\frac{7}{16}$
5. 18
6. £52

Review exercise

1. 3 : 2
2. 21 : 4
3. a $1 : \frac{1}{3}$ b 1 : 14   c $1 : \frac{1}{17}$ d $1 : 2\frac{2}{5}$
4. 5 : 2
5. a 0.5 m   b 16 m
6. 45 litres
7. 10
8. 110
9. Local professional (local $652.50, USA $684)
10. Small bottle (large 0.25 p/g, small 0.22 p/g)
11. 1.5 km²

Chapter 7 Answers

7.1 Get Ready

1. $4a$ 2. $8c$ 3. $2p^2$

Exercise 7A

1. a $7x + 4y$ b $10w + 2z$ c $4p + 5q$
   d $3a + b$ e $6c - 2d$ f $2m - 3n$
   g $4e - 7f$ h $2x + 10y + 2$ i $-2p + 3q - 5$
   j $13 - 5b - 4a$
2. $5x - 9$
3. $5x + 13y$

7.2 Get Ready

1. $4x + 2y + 12$ 2. $6y + 2x - 4$ 3. $8x + 2y + 6$

Exercise 7B

1. a 1   b 5   c 16   d 15
2. a -4 b -19 c 5   d 11
   e 17   f 11

7.3 Get Ready

1. 4^{11} 2. 7^7 3. 6^6

Exercise 7C

1. a m^5 b $6p^2$ c $20q^3$
2. a a^{11} b n^4 c x^6 d y^9
3. a $12p^6$ b $12a^5$ c $5b^9$ d $18n^3$
4. a $20t^8u^5$ b $6x^6y^7$ c $7a^5b^6$ d $8c^2d^9$
   e $24m^6n^5$

Exercise 7D

1. a a^3 b b^4 c c^3 d d
2. a $2q^2$ b $3p^5$ c $4x$ d $10y^7$
3. a $5a^2b^4$ b $5pq^3$ c $4c^2d^4$ d $3x^5$
   e $10m^2n$

Exercise 7E

1. a a^{14} b b^{15} c c^9 d d^{16}
2. a $4p^6$ b $81q^8$ c $25x^8$ d $\frac{m^{12}}{8}$
3. a $16x^{12}y^8$ b $49e^{10}f^6$ c $125p^{15}q^3$ d $\frac{8x^9}{27y^6}$

7.4 Get Ready

1. a^{18} 2. $27y^{15}$ 3. $\frac{4a^2}{b^6}$

Exercise 7F

1. a $\frac{1}{a}$ b $\frac{1}{b^2}$ c $\frac{1}{c^2}$ d $\frac{1}{d^3}$
2. a $\frac{1}{e^6}$ b $\frac{1}{f^8}$ c x^2 d y
3. a 1   b 1   c $\frac{1}{5p^2q^4}$
   d $\frac{1}{27c^9d^3}$ e $\frac{9r^4}{4p^6q^2}$

Exercise 7G

1. a $3a^2$ b $2c^{\frac{1}{2}}$ c $\frac{3e}{f^3}$ d $10x^{\frac{3}{2}}y^{\frac{5}{2}}$
2. a $\frac{1}{a^2}$ b $\frac{1}{2c}$ c $\frac{1}{2x^{\frac{9}{5}}y}$ d $\frac{1}{x^{\frac{1}{2}}y^{\frac{3}{2}}}$

7.5 Get Ready

1. 12, 14, 16   2. 34, 39, 44   3. 11, 13, 15

Exercise 7H

1. a add 3   b 14, 17   c 29
2. a add 6   b 20, 26   c 50
3. a subtract 7   b $-9, -16$ c -44
4. a add 1 to the difference of consecutive terms
   b 15, 21   c 55
5. a add 2 to the difference of consecutive terms
   b 20, 30   c 90

7.6 Get Ready

1. a add 3   b 13, 16   c 28
2. a add 3   b 11, 14   c 23
3. a subtract 6   b 94, 88   c 70

Exercise 7I

1. a i 2   ii -2
   b i 4   ii -11
   c i -5 ii 19
2. a $6n - 5$ b i 67   ii 295
3. a $4n + 3$ b i 63   ii 403
4. a $37 - 5n$ b i -63 ii -963
5. $7n + 11 = 103$ has no integer solution
6. The nth term in the sequence is $4n + 3$.
   $4n + 3 = 453$ has no integer solution.

Review exercise

1 **a** $5x - 5y$ **b** $6m - 10n$

2 **a** $240B + 114A$, B = British stamp, A = Australian stamp
b $375B + 212A$

3 **a** 3 **b** 8 **c** 7 **d** -30 **e** 62

4 **a** y^3 **b** $3x^2$ **c** z^8 **d** p^7 **e** $16a^7$

5 **a** a^3 **b** b^5 **c** $7p^3$ **d** $8x^3$ **e** $8a$

6 **a** subtract 3 **b** 87, 84 **c** 69

7 **a** 5 **b** -8

8 **a** $216 - 12n$ **b** **i** 60 **ii** -972

9 The nth term in the sequence is $3n + 2$. If $3n + 2 = 140$, $n = 46$. So 140 is the 46th term in the sequence.

10 **a** 55 cans
b Students' proofs
c 19 high (with 10 cans spare)

11 **a** $5\frac{1}{2}$ hours **b** 3.03 hours = 3 hours 2 mins
c Naismith's formula is for fit experienced walkers.

12 **a** a^{20} **b** $9b^8$ **c** $27e^{15}f^3$

13 Three consecutive even numbers are $2n$, $2n + 2$, $2n + 4$. Their sum $= 2n + 2n + 2 + 2n + 4 = 6n + 6$, which is always a multiple of 6.

14 $\dfrac{6x^2y}{4y^3} = \dfrac{3x^2}{2y^2}$. Squared numbers cannot be negative.

15 **a** $\dfrac{3p^2}{2y}$ **b** $\dfrac{1}{4q^{\frac{3}{2}}}$ **c** $\dfrac{2y}{x^2}$

16 64 cubes = 8 have 0 sides painted
24 have 1 side painted
24 have 2 sides painted
8 have 3 sides painted

Sides of cube	0	1	2	3
n by n by n	$(n - 2)^3$	$6(n - 2)^2$	$12(n - 2)$	8

Chapter 8 Answers

8.1 Get Ready

1 **a** $10x$ **b** $-12x^2$ **c** $2x^2$
d $x - 4$ **e** $x^2 + 5x + 6$
f $x^2 - 3x + 2$

Exercise 8A

1 **a** $2x + 6$ **b** $3p - 6$ **c** $4m + 4n$
d $15 - 3q$ **e** $4x + 2y - 6$ **f** $10c + 5$
g $4x^2 - 8$ **h** $3n^2 - 6n + 3$

2 **a** $y^2 + 2y$ **b** $g^2 - 3g$ **c** $2x^2 + 10x$
d $4n - n^2$ **e** $ab + ac$ **f** $3s^2 - 4s$
g $6t^2 + 3t$ **h** $4x^3 - 12x^2$

3 **a** $-2m - 6$ **b** $-6x - 6$ **c** $-m^2 - 5m$
d $-8y^2 - 12y$ **e** $-5p + 10$ **f** $-3q + 3q^2$
g $-2s^2 + 6s$ **h** $-12mn - 3n^2 + 15n$

Exercise 8B

1 **a** $8t - 3$ **b** $9p + 6$ **c** $11w + 6$
d $7d - 2$ **e** $5a + 3b$ **f** $5x + 3y + 5$

2 **a** $y + 20$ **b** $9a - 6$ **c** $-4x - 15$
d $q^2 - 3$ **e** $-5n$ **f** $11m^2 + 2m$

3 **a** $t - 16$ **b** $x + 19$ **c** $g^2 + g$
d $13c^2 - 22c$ **e** $4s^2 + 14s - 2$ **f** $p^2 + q^2$

4 **a** $3s - 4$ **b** $3m + 18$ **c** $5f^2 - 3f$
d $n^2 + 4n$ **e** $2x - x^2 + xy$ **f** $2p^2 + 5p$

8.2 Get Ready

1 **a** 2 **b** 5 **c** 4 **d** $3y$

Exercise 8C

1 **a** $3(x + 2)$ **b** $2(y - 1)$ **c** $5(p + 2q)$
d $7(2t - 1)$ **e** $2(4s + t)$ **f** $9(a + 2b)$
g $5(3u + v + 2w)$ **h** $t(x - y)$
i $c(a - 1)$ **j** $3(2x^2 + 3x + 1)$
k $2p(p - 1)$ **l** $q(q - 1)$ **m** $x(4x + 3)$
n $h(2 - 5h)$ **o** $p(p^2 + 2)$ **p** $s^2(1 + s)$

2 **a** $5x(y + t)$ **b** $3a(d - 2c)$ **c** $2p(3q + 2h)$
d $4y(2x - 1)$ **e** $2p(2q + s + 4t)$
f $mn(1 - k)$ **g** $2x(x + 2)$ **h** $12s(s - 2)$
i $2f^2(3 + f)$ **j** $y^2(y^2 + 1)$ **k** $cd(3d - 5c)$
l $ab(a^2 + b^2)$ **m** $2pr(4q + 5s)$ **n** $7ab(2a - b + 3)$
o $5x^2y(3 - 7y)$ **p** $3y(3y + 1)$

Exercise 8D

1 **a** $(x + 3)(x + 5)$ **b** $(x - y)(x + y)$
c $p(p + 1)$ **d** $(2t - s)(2t + s + 1)$
e $(a - 5)(a - 7)$ **f** $2(d + 1)(d + 1)$

2 **a** $2(y + 2)(y + 4)$ **b** $5(x - 1)(3x - 5)$
c $2(p + 5)(4p + 25)$ **d** $3(q + 1)(2q + 5)$
e $7(a + b)(a - b - 2)$ **f** $2x(x + 1)(2x - 3)$

8.3 Get Ready

1 24 cm^2 **2** $x \times (x + 2)$

Exercise 8E

1 **a** $x^2 + 7x + 12$ **b** $x^2 + 3x + 2$
c $x^2 - 3x - 10$ **d** $y^2 + y - 6$
e $y^2 - y - 2$ **f** $x^2 - 5x + 6$
g $a^2 - 9a + 20$ **h** $x^2 + 4x + 4$
i $p^2 + 8p + 16$ **j** $k^2 - 14k + 49$
k $a^2 + 2ab + b^2$ **l** $a^2 - 2ab + b^2$

2 **a** $2x^2 + 3x + 1$ **b** $3x^2 - 2x - 1$
c $2x^2 + 11x + 12$ **d** $3y^2 - 8y - 3$
e $2p^2 + 7p + 3$ **f** $6t^2 + 7t + 2$
g $6s^2 + 19s + 10$ **h** $4x^2 + 4x - 15$
i $12y^2 + 5y - 2$ **j** $6a^2 - 7a + 2$
k $9x^2 + 12x + 4$ **l** $4k^2 - 4k + 1$

3 **a** $x^2 + 3xy + 2y^2$ **b** $x^2 + xy - 2y^2$
c $x^2 - xy - 2y^2$ **d** $x^2 - 3xy + 2y^2$
e $6p^2 + 7pq - 3q^2$ **f** $6s^2 - 7st + 2t^2$
g $4a^2 + 12ab + 9b^2$ **h** $4a^2 - 12ab + 9b^2$

8.4 Get Ready

1 **a** 1 and -6, 6 and -1, 2 and -3, 3 and -2
b 1 and 15, -1 and -15, 3 and 5, -3 and -5

2 5 and 2

3 -3 and -5

Exercise 8F

1 a $3, 5$ b $-6, -4$ c $-6, -3$
   d $4, -2$ e $-4, 2$ f $-3, 3$
2 a $(x + 3)(x + 5)$ b $(x + 1)(x + 7)$ c $(x + 4)(x + 5)$
   d $(x - 5)(x - 1)$ e $(x - 8)(x - 1)$ f $(x - 1)^2$
   g $(x - 3)(x + 6)$ h $(x - 6)(x + 3)$ i $(x - 4)(x + 7)$
   j $(x - 4)(x + 3)$ k $(x - 4)(x + 6)$ l $(x - 2)(x + 2)$
   m $(x - 9)(x + 9)$

Exercise 8G

1 a $(x - 6)(x + 6)$ b $(x - 7)(x + 7)$
   c $(y - 12)(y + 12)$ d $(5 - y)(5 + y)$
   e $(w - 50)(w + 50)$ f $(100 - a)(100 + a)$
   g $(x - 1)(x + 3)$ h $y(18 - y)$
   i $4ab$
2 a 2800 b 50 c 0.75 d $20\,000$
3 a $(2x - 7)(2x + 7)$ b $(3y - 1)(3y + 1)$
   c $(11t - 20)(11t + 20)$ d $-(q + 1)(q + 3)$
   e $8t$ f $4(p + q)$
   g $4(5p - q + 2)(5p + q + 3)$
   h $100st$
4 a $3(x - 2)(x + 2)$ b $5(y - 5)(y + 5)$
   c $10(w - 10)(w + 10)$ d $4(p - 4q)(p + 4q)$
   e $3(2a - 3b)(2a + 3b)$ f $8x$

Exercise 8H

1 a $(5x + 1)(x + 3)$ b $(2x + 1)(x + 5)$
   c $(3x + 1)(x + 1)$ d $(4x + 1)(2x + 1)$
   e $(3x + 2)(2x + 3)$ f $(6x - 1)(x - 1)$
   g $(5x - 2)(x - 1)$ h $(4x - 1)(3x - 2)$
   i $(4x + 3)(2x - 1)$ j $(2x + 3)(x - 5)$
   k $(7x + 2)(x - 3)$ l $(3x + 2)(x - 4)$
   m $(2y + 1)(2y + 5)$ n $(6y - 1)(y - 2)$
   o $(3y - 5)(2y - 5)$
2 a $2(3x + 4)(x + 1)$ b $3(2y - 1)(y - 2)$
   c $5(x + 2)(x - 1)$
3 a $(x - y)(x + 2y)$ b $(x + y)(2x + 5y)$
   c $(3x - 2y)(2x + 3y)$

Review exercise

1   $5x - 2$
2 a $5(m + 2)$ b $y(y - 3)$
3 a $(a + b)(x + y)$ b $(a - b)(c + d)$
4   $x^2 - x - 12$
5 a $a^2 + 4a + 4$ b $c^2 - 6c + 9$
   c $d^2 + 2d + 1$ d $x^2 + 2xy + y$
6 a $x^2 + 15x + 50$ b $y^2 + 18y + 81$
   c $x^2 - 2x - 8$ d $x^2 - x - 6$
   e $t^2 - 7t + 6$ f $2x^2 + 11x + 12$
   g $6p^2 + p - 1$ h $4c^2 - d^2$
   i $16y^2 - 8y + 1$
7 a $(t + 5)(t + 6)$ b $(x + 7)^2$ c $(p + 5)(p - 3)$
   d $(y - 6)^2$ e $(x - 4)(x - 1)$ f $(s - 8)(s + 8)$
8 a $(x + 6)(x + 7)$ b 11×17
9   $2(Y + 3)$, 26 red flowers
10 a $(x - 20)(x + 20)$ b $(3t - 2)(3t + 2$
   c $(10 - y)(10 + y)$ d $(5 - 2p)(5 + 2p)$
11 a 41 b 1.99 c 16

12 8000
13 Three consecutive numbers are $n, n + 1, n + 2$.
    $(n + 1)(n + 2) - n(n + 1) = (n^2 + 3n + 2) - (n^2 + 1)$
    $= 2n + 2 = 2(n + 1)$
14 $6(x + 2)$
15 a Each team plays 3 games at home against the other
      teams.
      So total number of games $= 4 \times 3 = 12$
   b 380 c $a^2 - a = a(a - 1)$
16 a $(2x + 1)(x + 2)$ b $(2w - 1)(w + 3)$
   c $(3a + 2)(a + 4)$ d $(3z - 2)(10z - 1)$
   e $(8y - 1)(y + 3)$ f $(3p + q)(2p - q)$
17 Let the top left number in the 2 by 2 square be n.

n	$n + 1$
$n + 6$	$n + 7$

Difference of products from opposite corners
$= (n + 1)(n + 6) - n(n + 7) = (n^2 + 7n + 6) - (n^2 + 7n)$
$= 6$

Chapter 9 Answers

9.1 Get Ready

1 a $x = 7$ b $x = 1$
2 a $y = 1$ b $y = 1.5$

Exercise 9A

1 a

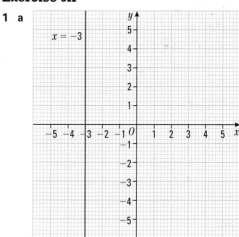

   b

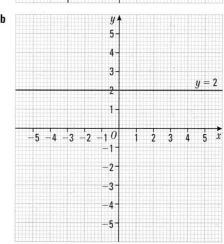

Answers

c

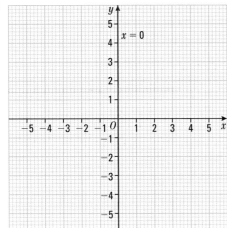

f

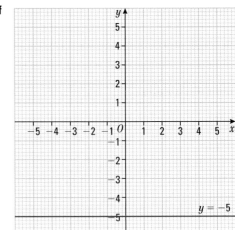

2 **a** $x = -4$ **b** $y = 2$ **c** $y = 5$ **d** $x = -\frac{1}{2}$

3 **a** $(1, 3)$ **b** $(-4, 2)$ **c** $(-\frac{1}{2}, 3)$

4 22 units, 28 units squared

Exercise 9B

1 a

x	-2	-1	0	1	2	3	4
y	8	6	4	2	0	-2	-4

d

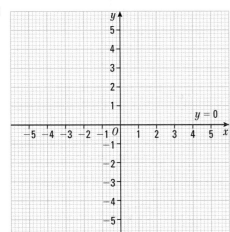

b

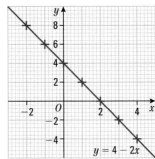

e

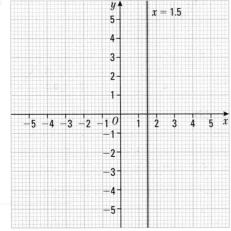

2 a

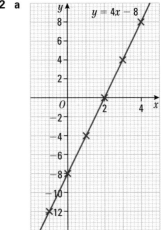

b **i** $(2, 0)$ **ii** $(0, -8)$
c **i** 2 **ii** 3.5

3 a

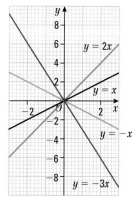

b all pass through (0, 0)

4 a

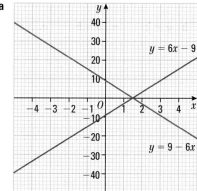

b (1.5, 0)

5

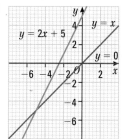

6.25 square units

Exercise 9C

1

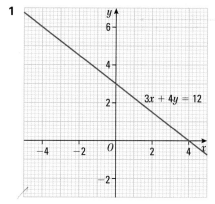

2 a

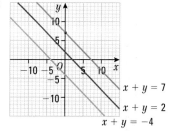

b They are parallel to each other.

3

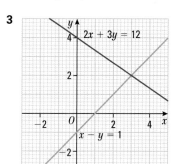

(3, 2)

4 a

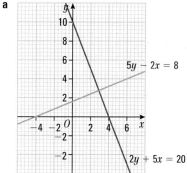

b They intersect at right angles.
c e.g. $2x + 3y = 6$ and $3x - 2y = 12$

9.2 Get Ready

1 a 5   **b** 5.5   **c** 17.25
   d 0.225   **e** 1   **f** −8.5

Exercise 9D

1 a (1, 2)   **b** (3, −0.5)   **c** (−1.5, 3)
   d (−2.5, −0.5)   **e** (−4, 0.5)   **f** (−0.5, −2)
   g (2.5, −0.5)   **h** (0.5, 2.5)   **i** (−2, 3)
   j (4.5, −0.5)
2 a (1.5, 0.5)   **b** (4, 1)   **c** (1.5, 1)
   d (3.5, 2.5)
3 a (4, 4)   **b** (−2, 2.5)   **c** (1.5, −3.5)
   d (−3, 4.5)   **e** (2, −2.5)   **f** (2.5, −1.5)

9.3 Get Ready

1

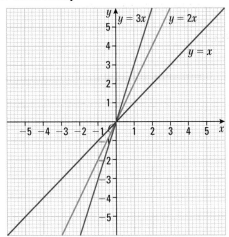

2

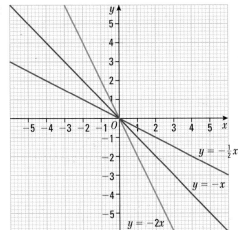

3

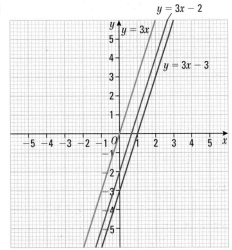

3 a

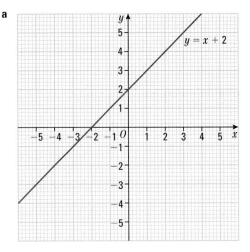

b

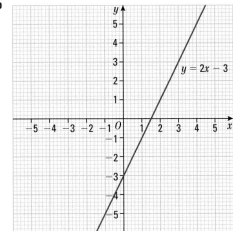

c

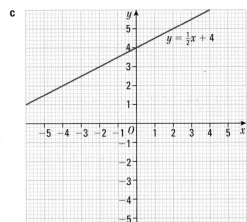

Exercise 9E

1 a 30   **b** 10   **c** $\frac{10}{3}$

  d $-\frac{2}{5}$ **e** $-\frac{1}{5}$ **f** $-\frac{1}{40}$

2 a 2   **b** -2

  c 4   **d** $\frac{3}{4}$

d

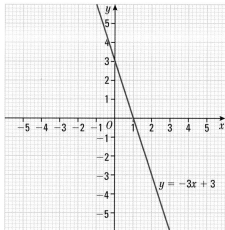

$y = -3x + 3$

e

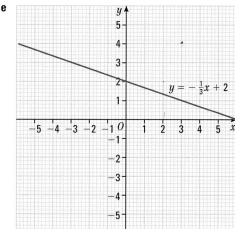

$y = -\frac{1}{3}x + 2$

4 e.g. (3, 2)

5 a $-\frac{1}{2}$ **b** (0, 4)

Exercise 9F

1 a Gradient = 40. This represents the extra time needed (in minutes) for each extra kilogram of chicken.

b Cooking time = 40 minutes per kilogram plus 20 minutes

c You can't have a negative weight of chicken.

2 a 50°F

b Gradient = 1.8. This represents the number of degrees Fahrenheit for each degree Celsius.

3 a A 15, B 10, C 4

b The gradients represent speed. A is the car (the fastest vehicle), B is the lorry, C is the cycle (the slowest vehicle).

4 a

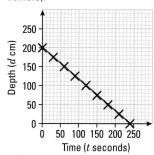

b $-\frac{5}{6}$

c The depth of water is going down by $\frac{5}{6}$ cm per second. The swimming pool is being emptied.

9.4 Get Ready

1 $2, \frac{1}{2}, -1$
The gradient is the same as the coefficient of x.

2 $-1, 2, 2$
The y-intercept is the same as the value of the number in the equation.

Exercise 9G

1 $y = 2x + 5$

2 a i 4   **ii** 1

  b i 3   **ii** -4

  c i $\frac{2}{3}$ **ii** 4

  d i -0.4 **ii** (0, 4)

  e i $1\frac{1}{3}$ **ii** (0, -4)

  f i $\frac{1}{2}$ **ii** (0, 0)

3 A $y = 2x + 8$ **B** $y = \frac{1}{3}x + 2$ **C** $y = 5 - \frac{1}{2}x$
    D $y = 4 - x$ **E** $y = -2x - 6$

4 $y = 5x - 2$

5 $y = 3x - 10$

9.5 Get Ready

1 a $-\frac{1}{2}, 2$ **b** $-\frac{1}{3}, 3$ **c** $\frac{2}{3}, \frac{2}{3}$

In **a** and **b** the product of the gradients is -1.
In **c** the gradients of the parallel lines are the same.

Exercise 9H

1 a $-\frac{1}{3}$ **b** $\frac{1}{4}$ **c** -5 **d** $-\frac{1}{3}$ **e** 6

2 a $y = 2x + c$ for any value of c except 5

  b $y = \frac{1}{3}x + c$ for any value of c except -1

  c $y = c - x$ for any value of c except 4

3 a $y = c - x$ for any value of c

  b $y = c - \frac{1}{3}x$ for any value of c

  c $y = 2x + c$ for any value of c

4 $y = 4x + 3$

5 $2x + y = 0$

6 $y = -4x$

7 $y = x - 3$

9.6 Get Ready

1 C 14:30, D 15:15, E 15:54

Exercise 9I

1 a 4 km   **b** 75 minutes   **c** $5\frac{1}{3}$ km/h

  d 6 minutes each time   **e** 10 km/h

2 a £2.75   **b** 2 kg   **c** £12.20

3 a   **b**

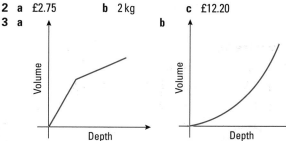

Answers

c

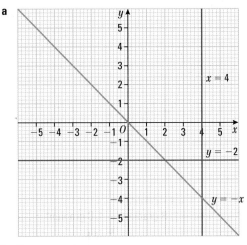

d

4 A **b**, B **a**, C **c**

5

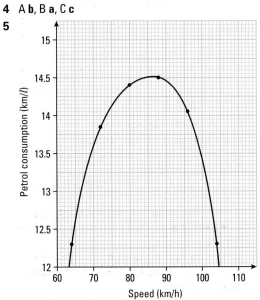

a 13.5 km/*l*   **b** 74 km/h, 97 km/h

9.7 Get Ready

1. 60 minutes
2. $\frac{1}{4}$
3. 0.6
4. 18 minutes
5. 5 hours 42 minutes
6. 7 hours 42 minutes

Exercise 9J

1. **a** 13 km/litre   **b** 6 litres
2. **a** 240°   **b** 30 seconds
3. **a** 60 litres/min   **b** 18 minutes 30 seconds
4. 0.0625 litre

9.8 Get Ready

1. **a** 18 km/*l*   **b** 6.5 km/*l*   **c** 13.2 km/*l*

Exercise 9K

1. 4.24 km/h   **2** 10 km/h
3. 1 hour 12 minutes   **4** 306 km/h
5. The speed for the 100 m was 10.32 m/s. The speed for the 200 m was 10.36 m/s. The 200 m race was won with the faster average speed.

Review exercise

1 a

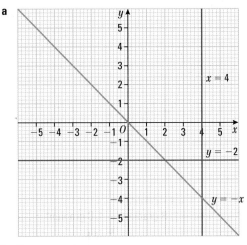

b 2 square units

2

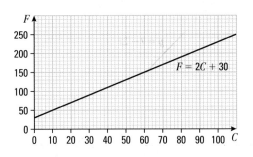

3 a i 3   **ii** $-\frac{1}{2}$ **iii** 0
   b i $\frac{1}{2}$ **ii** -3

4

C	0	20	40	60	80	100
F	30	70	110	150	190	230

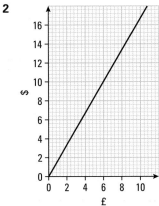

$F = 2C + 30$

5 a

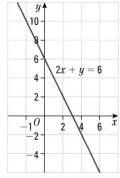

b i −2 **ii** 6

6 a 120 km **b** 0.5 hour **c** 80 km/h
 d 18:52 **e** 48 km/h

7 B (A gradient = 2, B gradient = 4, C gradient = 2.5)

8 a

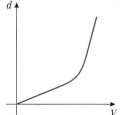

b

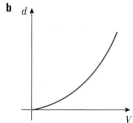

c

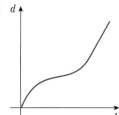

d

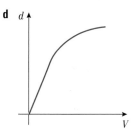

9 a No **b** $y = \frac{1}{2}x + 5$ **c** $y = -2x + 9$

10

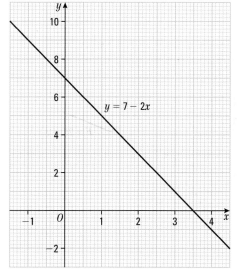

11 a 4 **b** $\left(-\frac{3}{4}, 0\right)$ **c** 4.5

12 a From the top on the right-hand side: B, A, C
 b £40
 c C would be cheapest. (A £32, B £30, C £25)

13 If Abbie plans to go to the health club more than 50 times a year, she should choose Atlantis.

14 (3, 2)

15 80 km/hour

16 432 miles/hour

17 John's speed: 20 km/hour. Kamala's speed: 21 km/hour. Kamala had the greater average speed.

18 320 seconds

19

Equation of line	Gradient	y-intercept
$y = 2x + 5$	2	5
$y = 7x - 3$	7	−3
$y = 6 - x$	−1	6
$y = \frac{2}{3}x - 1$	$\frac{2}{3}$	−1
$y = 3 - x$	−4	3

20 A: The temperature stays constant.
 B: The temperature rises at a constant rate.
 C: The temperature rises at a constant rate and then falls at a faster constant rate.
 D: The temperature stays the same and then falls at a constant rate.
 E: The temperature rises at a constant rate, stays the same for a period of time and then continues to rise at the same constant rate.
 F: The temperature rises at a constant rate, stays the same for a period of time and then falls at the same rate at which it rose.

21 a Perimeter = 2 × (2x + y) = 24, so 2x + y = 12
 b

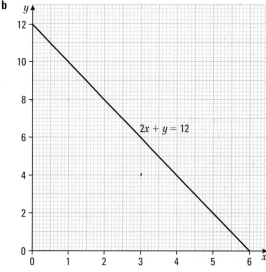

 c $x = 3$

22 $k = 7$. $x = 3$ in $y = 3x - 2$ gives $y = 7$, so (3, 7) also lies on $y = 3x - 2$

23 $y = 5 - 2x$

24 a $-\frac{1}{2}$ **b** $y = 2x - 5$

Chapter 10 Answers

10.1 Get Ready

1 a 6 **b** 2 **c** 4

Answers

Exercise 10A

1 formula   2 identity
3 expression   4 equation
5 formula   6 formula
7 expression   8 equation
9 formula   10 equation
11 equation   12 identity

10.2 Get Ready

1 a 5347.76   b $\frac{8}{3}$, or $2\frac{2}{3}$

Exercise 10B

1 a 120   b -25
2 a 65   b 5
3 a 1570   b 120.576
4 a 120 miles   b 8 m/s
5 a 8 hours 50 minutes   b 18 kg
6 a 12   b 9

10.3 Get Ready

a 3200   b 9

Exercise 10C

1 a $C = 50d + 90$ b €790
2 $P = 5x + 3y + z$
3 $T = \dfrac{45m + 75f}{60} = \dfrac{3m + 5f}{4}$
4 $P = 4x + 4y$
5 a $T = 2a + 5c$ b $P = 2a + 3c$

Review exercise

1 a IDENTITY   b EXPRESSION
   c FORMULA   d EXPRESSION
   e EQUATION   f IDENTITY
2 a 51   b -3
3 $2c + 4r$
4 a $b - 3$ b $2b$
5 a -13 b 15
6 a 1.71 or $\frac{12}{7}$ or $1\frac{5}{7}$ b -20
7 9
8 a $b = 3a$ b $c = a + 5$
9 a £130   b $C = 90 + 0.5m$
10 0.6
11 a 166.25   b 4

Chapter 11 Answers

11.1 Get Ready

1 a $(x + 4)(x + 1)$
   b $(x - 3)(x + 2)$
   c $(2x + 1)(x + 3)$

Exercise 11A

1 a $2x^3$ b $\dfrac{x}{3y}$ c $\dfrac{x - 5}{2}$ d x e $-2x$

2 a $\dfrac{x + 1}{x + 2}$ b $\dfrac{x + 1}{x}$ c $\dfrac{x - 2}{x + 4}$ d $\dfrac{x - 4}{x + 3}$

3 a $\dfrac{x + 1}{x}$ b $\dfrac{4x}{x - 6}$ c $\dfrac{2(x - 2)}{x + 2}$ d $\dfrac{x - 3}{x}$

4 a $\dfrac{2x + 3}{3x + 2}$ b $\dfrac{5x - 3}{3x - 2}$ c $\dfrac{3x + 1}{3x - 1}$ d $\dfrac{x + 1}{2x + 3}$

5 a $\dfrac{x + 5}{x - 5}$ b $\dfrac{x - 5}{2(x + 5)}$ c $\dfrac{2x - 1}{2x}$

   d $\dfrac{3 - x}{x + 3}$ e $\dfrac{2 - x}{x + 2}$ f $-(x + 4)$

11.2 Get Ready

1 a 30   b $12x$ c $x(x + 1)$

Exercise 11B

1 a x b $2x$ c $\dfrac{7}{10x}$ d $\dfrac{4x}{9}$

   e $\dfrac{x}{5}$ f $\dfrac{5}{3x}$

2 a $\dfrac{7x}{12}$ b $\dfrac{x}{3}$ c $\dfrac{3x}{8}$ d $\dfrac{5x}{6}$

   e $\dfrac{5}{6x}$ f $\dfrac{1}{4x}$

3 a $\dfrac{5x + 2}{6}$ b $\dfrac{9x - 7}{20}$ c $\dfrac{x}{9}$

   d $\dfrac{2x + 5}{(x + 2)(x + 3)}$ e $\dfrac{x - 2}{(x + 2)(x + 1)}$

   f $\dfrac{4}{(2x - 1)(2x + 3)}$

Exercise 11C

1 a i $2(x + 1)$ ii $6(x + 1)$
   b $6(x + 1)$ c $\dfrac{2}{3(x + 1)}$

2 a $15x$ b $(x + 2)(x + 3)$
   c $x(x - 1)$ d $(x + 1)(x + 2)$
   e $2(x - 3)$ f $x(x + 1)$

3 a $(x + 1)(x + 2)$ b $\dfrac{x}{(x + 1)(x + 2)}$

4 a $(x - 2)(x + 2)$ b $\dfrac{3x + 4}{(x - 2)(x + 2)}$

5 a $(2x - 1)(x - 1)$ b $\dfrac{1}{(x - 1)}$

6 $\dfrac{x - 1}{2(x + 1)(x + 3)}$

7 $\dfrac{2x^2 + 7x + 4}{8x(x + 1)}$

8 $\dfrac{3x}{(3 - x)(3 + x)}$

9 a i $(x + 4)(x + 5)$ ii $(x + 5)(x + 6)$
   b $\dfrac{3x + 20}{(x + 4)(x + 5)(x + 6)}$

10 $\dfrac{4}{(2x - 1)(2x + 1)(2x - 3)}$

11.3 Get Ready

1 a a b $\dfrac{1}{2}$ c $\dfrac{x + 3}{x + 1}$

Exercise 11D

1 a $\dfrac{x^2}{15}$ b $\dfrac{12}{y^2}$ c $\dfrac{15xy}{8}$ d $\dfrac{x(x - 3)}{12}$

2 a $\dfrac{xy}{2}$ **b** $\dfrac{2x}{9}$ **c** $\dfrac{6}{y}$ **d** $\dfrac{2(x+1)}{x-1}$

3 a $\dfrac{5}{9}$ **b** $\dfrac{5xy}{18}$ **c** x^2y^2 **d** $\dfrac{2x(x+2)}{(x+1)^2}$

4 a $\dfrac{27}{4}$ **b** $\dfrac{5x}{8}$ **c** $\dfrac{7x}{2y^2}$ **d** $\dfrac{2x}{x-5}$

5 a $\dfrac{(x+1)^2}{2}$ **b** $(x+2)(x-1)$ **c** $\dfrac{x}{x+2}$

  d $\dfrac{1}{6}$ **e** $3(3x-1)$ **f** $\dfrac{3(x+4)}{4}$

6 a $(x-2)(x+2)$ **b** $\dfrac{x-2}{x^2+4}$

7 a i $(x+1)(x+4)$ **ii** $(x+2)(x+4)$

  b $\dfrac{(x+2)(x+3)}{(x+1)^2}$

8 $x+1$

11.4 Get Ready

1 a Even **b** Either **c** Odd **d** Either
  e Either

Exercise 11E

1 $(2n-1)+2m = 2(m+n)-1$

2 $\frac{1}{2}[n+(n+1)+(n+2)+(n+3)] = \frac{1}{2}[4n+6] = 2n+3$

3 $n+(n+1)+(n+2) = 3n+3 = 3(n+1)$

4 a $(2n-1)2m = 2[m(2n-1)]$
  b $(2n-1)(2m-1) = 4mn-2m-2n+1$
    $= 2(2mn-m-n)+1$
  c $(2n)(2m) = 4mn = 2(2mn)$

5 $(m-n)(m+n) = m^2-nm+mn-n^2 = m^2-n^2$

6 $(n+4)^2-n^2 = n^2+8n+16-n^2 = 8n+16 = 8(n+2)$

Review exercise

1 $n+(n+1)+(n+2)+(n+3) = 4n+6$
  $(n+3)(n+2)-n(n+1) = (n^2+5n+6)-(n^2+n)$
  $= 4n+6$

2 a $\dfrac{x^2}{2}$ **b** $\dfrac{(x+1)^2}{3}$ **c** x **d** $\dfrac{x+1}{x+2}$

  e $\dfrac{x}{x+3}$ **f** $\dfrac{x-5}{x+5}$ **g** $\dfrac{x-1}{x(x+1)}$ **h** $\dfrac{x+4}{3x+2}$

3 a $\dfrac{x}{2}$ **b** $\dfrac{13}{6x}$ **c** $\dfrac{10x}{(5x-3)(5x+3)}$

  d $\dfrac{2x+7}{(x+1)(x+2)}$ **e** $\dfrac{x+2}{(x+1)^2}$

  f $\dfrac{2}{(x+1)(x+5)}$

4 a 2 **b** x **c** $\dfrac{2}{5}$ **d** $\dfrac{2s^2}{3}$

  e $\dfrac{x-1}{(x+1)(x+3)}$ **f** 1 **g** $\dfrac{2}{3}$ **h** $\dfrac{2x+1}{x}$

5 a $(n+1)(n+2)+n(n+1) = n^2+3n+2+n^2+n$
    $= 2n^2+4n+2$
    $= 2(n^2+2n+1)$
    $= 2(n+1)^2$
  b $2(n+1)^2$ is always even.

6 $(2n+1)^2-(2n-1)^2 = (4n^2+4n+1)-(4n^2-4n+1)$
    $= 8n$

7 $\dfrac{100-(x^2-16x+64)}{4} = \dfrac{36+16x-x^2}{4} = \dfrac{(2+x)(18-x)}{4}$

8 a $4-2=2, 6-4=2, 8-6=2, \ldots$
  The difference between each pair of consecutive even numbers is 2. Therefore if the nth even number is $2n$, the next even number must be $2n+2$.
  b $2n+(2n+2)+(2n+4) = 6n+6 = 6(n+1)$, which must be a multiple of 6

9 $(3n+1)^2-(3n-1)^2 = (9n^2+6n+1)-(9n^2-6n+1)$
  $= 12n = 4 \times 3n$, which must be a multiple of 4

10 $\dfrac{16(2x-1)}{x(x-1)}$ hours

Chapter 12 Answers

12.1 Get Ready

When you fold it over, the star fits exactly on top of itself. When you rotate the star, it fits exactly on top of itself in four different positions.

Exercise 12A

1 a Yes

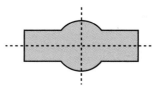

  b Yes

  c No **d** No **e** Yes

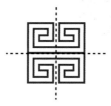

  f No

2 a No rotational symmetry
  b Rotational symmetry of order 2
  c Rotational symmetry of order 3
  d Rotational symmetry of order 8
  e Rotational symmetry of order 2
  f No rotational symmetry

3 a

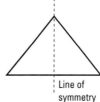

Line of symmetry

  b Isosceles triangle

4 a

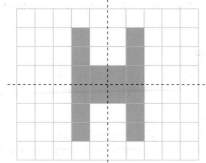

b

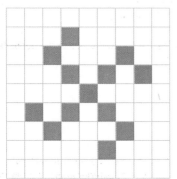

5 a e.g. 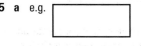   **b** e.g.

c e.g.

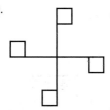

12.2 Get Ready

1 a i An isosceles triangles has two sides the same length and two angles the same.
  ii An equilateral triangle has all three sides the same length and all three angles the same.
b Yes
2 A polygon with four sides.

Exercise 12B

1 a

b It is also an isosceles triangle.
2 a No. It could be a rectangle, square, rhombus, parallelogram, trapezium or isosceles trapezium.
  b No. It could be a rectangle, rhombus or parallelogram.

c No. It could be a rectangle or rhombus.
d Yes. It is a rectangle.

3 e.g.

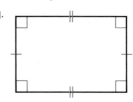

4 e.g.

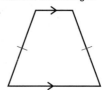

Rotational symmetry of order 2

12.3 Get Ready

1 a 150 g   **b** 3 cm   **c** 5 m   **d** 300 ml

Exercise 12C

1 a	600 cm	**b**	21 cm	**c**	510 cm		
d	84 cm	**e**	5.9 cm	**f**	48.3 cm		
g	300 000 cm	**h**	6700 cm				
2 a	3000 kg	**b**	8200 kg	**c**	6 kg		
d	0.9 kg	**e**	0.43 kg	**f**	4.7 kg		
3 a	2 litres	**b**	7 litres	**c**	5.9 litres		
d	45 litres						

4 7 litres

Exercise 12D

1 8.8 pounds
2 50 kg
3 4 litres
4 8 gallons
5 17.5 pints
6 360 cm
7 60 miles
8 96 km
9 585p per gallon

Exercise 12E

1 a 80 ounces   **b** 6 pounds
2 180 inches
3 a 132 pounds   **b** 60 kg

Review exercise

1 a

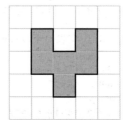

b

(grid shape)

2 **a** 110 m, 70 g, 40 litres **b** 400 cm **c** 1.5 kg
3 No, 1.5 km is 1500 m.
4 **a** **i** metres
          ii kilograms
      b 20 mm
5 **a** **i** kilometres
          ii litres
      b **i** 50 mm
          ii 4 kg
6 **a** 3 **b** 3 **c** he can use any corner
7 **a** D **b** B **c** A
8 **a** D, A **b** B, C
9 **a**

(T-shape figure with dashed vertical line of symmetry)

      b 2

10 No, Carla could have as little as 1.5 litres of water. The minimum amount of water Gurpia can have is 1949.5 ml.

Chapter 13 Answers

13.1 Get Ready

1 $a = 135°, b = 45°, c = 135°$ **2** 40° **3** 48°

Exercise 13A

1 $a = 63°$ (corresponding angles)
    $b = 49°$ (corresponding angles)
    $c = 68°$ (angles in a triangle add up to 180°)
2 $p = 113°$ (corresponding angles and angles on a straight line)
    $q = 67°$ (corresponding angles)
    $r = 113°$ (alternate angles or angles on a straight line)
3 $l = 81°$ (vertically opposite angles)
    $m = 54°$ (alternate angles)
    $n = 45°$ (angles of a triangle add up to 180°)
4 $y = 58°$ (alternate angles)
    $z = 58°$ (alternate angles and angles on a straight line)
5 $g = 57°$ (isosceles triangle and alternate angles)
    $h = 180 − 2 × 57 = 66°$ (angles of a triangle add up to 180° and alternate angles)
    $k = 114°$ (angles of a triangle add up to 180° and angles on a straight line)
6 $a = 50°$ (2 sets of alternate angles)
7 **a** a and p, b and q, c and s, d and r
    b a and r, b and s, c and q, d and p
    c a and b, b and d, d and c, c and a, p and q, q and r, r and s, s and p
        The angles are on a straight line.

8 Angle BAC = Angle DCE = 56°, so they are corresponding angles

13.2 Get Ready

1 $a = 50°, b = 80°$ **2** $c = 28°, d = 28°$
3 $e = 60°$

Exercise 13B

1 angles in a triangle, angles on a straight line
2 same angle, same angle, angles in a triangle, angles in a triangle
3 **a** $b = d$ (alternate angles)
        $a = c$ (corresponding angles)
        so $a + b = c + d$
    b The exterior angle is equal to the sum of the opposite two interior angles.

13.3 Get Ready

$j = 143°$

Exercise 13C

1 141° (angle sum of quadrilateral / equilateral triangle)
2 126° (angles on a straight line / exterior angle of a triangle)
3 **a** $a = 132°$ (symmetry), $b = 37°$ (angle sum of quadrilateral)
    b 66°, 114°, 114° (symmetry / angle sum of quadrilateral)
4 113° (vertically opposite angles / angles on a straight line / angles at a point / angle sum of a quadrilateral)

13.4 Get Ready

1 **a** alternate angles **b** 90°

Exercise 13D

1 alternate angles
2

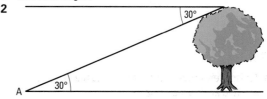

13.5 Get Ready

$a = 43°$ (alternate angles)
$b = 72°$ (opposite angles)
$c = 65°$ (angles sum of a triangle)
$d = 64°$ (angles on a straight line)
$e = 58°$ (angle sum of an isosceles triangle)
$f = 54°$ (opposite angles of a parallelogram)
$g = 126°$ (angles at the end of a parallelogram)

Exercise 13E

1 **a** 124° (isosceles triangle / angle sum of a triangle / vertically opposite angles)
    b 56° (angles on a straight line / corresponding angles)
2 L = 70°, M = 55°, N = 55° (alternate angles / angles on a straight line)

Answers

3 a $p = 57°$ (exterior angle of a triangle / angle sum of a quadrilateral)
  b $q = 117°$ (exterior angle of a triangle / corresponding angles)
4 a alternate angles
  b $b = d$ (alternate angles)
    $a = c$, part **a**
    so $a + b = c + d$
  c Opposite angles in a parallelogram are equal.
5 $a + b + c + d = 360°$ (angles in a quadrilateral)
  $a + c = 180°$ (given in question)
  so $b + d = 180°$

13.6 Get Ready

1 Equilateral triangle
2 A square is a quadrilateral with **equal** sides and **equal** angles.

Exercise 13F

1 a

Polygon	Number of sides (n)	Number of diagonals from one vertex	Number of triangles formed	Sum of interior angles
Triangle	3	0	1	180°
Quadrilateral	4	1	2	360°
Pentagon	5	2	3	540°
Hexagon	6	3	4	720°
Heptagon	7	4	5	900°
Octagon	8	5	6	1080°
Nonagon	9	6	7	1260°
Decagon	10	7	8	1440°

  b i $n - 3$ **ii** $n - 2$ **iii** $(n - 2) \times 180°$
2 Angles not equal

Exercise 13G

1 a 15   **b** 18   **c** 160°
2 a 120°   **b** 144°   **c** 168°
3 a 142°   **b** 103°
4 360 is not divisible by 25
5 a $a = 32°, b = 30°, c = 42°, d = 63°, e = 44°, f = 27°,$
    $g = 59°, h = 63°$
  b 360°
6 a 135° (angle sum of isosceles triangle / angles on a straight line)
  b i The interior angles are all the same.
    ii Not all the sides are the same length.

Exercise 13H

1 a i 110°   **ii** 143°
  b i 50°   **ii** 36°
2 a 72°   **b** 45°   **c** 30°   **d** 14.4°
3 a 24   **b** 3960°

4 144°, 98°, 129°, 128°, 107° and 114°
5 Exterior angle = 12°, interior angle = 168°, angle BCA = 6° (angle sum of isosceles triangle)
6 Angle BCD = $180 - e$ (angles on a straight line)
  So angle BCO = $(180 - e) \div 2$ (by symmetry)
  And angle CBO = $(180 - e) \div 2$ (isosceles triangle)
  So $c = 180 - \{(180 - e) \div 2\} - \{(180 - e) \div 2\}$ (angles in a triangle) giving $c = e$

13.7 Get Ready

1 a 20°   **b** 65°   **c** 40°

Exercise 13I

1 $a = 25°$
2 $b = 33°$
3 $c = 40°$
4 $d = 19°$ (angles on a line, isosceles triangle in a circle)
5 $e = 60°, f = 40°, g = 100°$ (equilateral triangle, isosceles triangle in a circle)
6 $h = 54°, i = 126°, j = 27°$ (right-angled triangle, angles on a line, isosceles triangle in a circle)

13.8 Get Ready

1 $a = 33°, b = 57°, c = 90°$

Exercise 13J

1 $a = 50°$ **2** $b = 59°$
3 $c = 136°$ **4** $d = 66°, e = 24°$
5 $f = 60°$ **6** $g = 107°$

Review exercise

1 $88 + 96 \neq 180$, so the lines are not parallel. Ben is right.
2 a 44°   **b** 44°
3 56°
4 a i $x = \dfrac{180 - 54}{2} = 63°$
    ii angle sum of an isosceles triangle
  b $y = 54 + 63 = 117°$ (exterior angle of a triangle)
    or
    $y = 180 - 63 = 117°$ (angles on a straight line)
5 47°
6 i 76°
  ii Line TO and SO are equal and perpendicular to PT and PS respectively.
    As OTS = OST we know PTS = STP = 50°.
    As PTS is a triangle $x = 180 - 52 - 52 = 76$.
7 123°
8 $x = 130°$ (angles on a straight line)
  $y = 50°$ (alternate angles)
9 $x = 180 - (360 - 50 - 119 - 105) = 94°$ (angle sum of a quadrilateral / angles on a straight line)
10 angle ABQ = 90° (angle in a square)
  angle ABC = $180 - \dfrac{360}{6} = 120°$ (exterior angle of a regular polygon / angles on a straight line)
  $x = 360 - 90 - 120 = 150°$ (angles around a point)

11 angle in equilateral triangle $= 60°$

base angle in isosceles triangle $= \dfrac{180 - 57}{2} = 61.5°$

(angle sum of a triangle)

$p = 360 - 60 - 61.5 = 238.5°$ (angles around a point)

12 a i $w = 25°$ (base angles of an isosceles triangle)

ii $x = 180 - 2 \times 25 = 130°$ (angle sum of a triangle)

b angle SQR $= 180 - 130 = 50°$ (angles on a straight line)

$y = \dfrac{180 - 50}{2} = 65°$ (angle sum of an isosceles triangle)

13 exterior angle $= \dfrac{360}{10} = 36°$

$x = 180 - 36 = 144°$ (angles on a straight line)

14 a $x = \dfrac{180 - 120}{2} = 30°$ (angle sum of an isosceles triangle)

b angle ABD $= 180 - 30 = 150°$ (angles on a straight line)

$y = 360 - 150 - 54 - 108 = 48°$ (angle sum of a quadrilateral)

15 angle ACB $=$ angle ABC $= x + 20$ (base angles of an isosceles triangle)

angle BAC $+ 2(x + 20) = 180°$ (angle sum of a triangle)

angle BAC $= 140 - 2x$

16 a interior angle $= 180 -$ exterior angle

(angles on a straight line)

interior angle $= 180 - \frac{2}{3} \times$ interior angle

interior angle $= 108°$

b exterior angle $= 72° = \dfrac{360}{n}$

$n = 5$

17 Angle $CAB = 360 - x - 90 - 80 - x$

(Angles at a point add up to 360°)

$= 190 - 2x =$ angle ABC

(Base angles in an isosceles triangle are equal)

Angle $BCD = 190 - 2x + 190 - 2x$

(exterior angle of a triangle $=$ sum of two interior opposite angles)

$= 380 - 4x = 4(95° - x)$

18 Angle $PQR =$ angle $PRQ = (180° - 20°) \div 2 = 80°$

(base angles of an isosceles triangle are equal and Angles in a triangle add up to 180°)

Angle $PQY = 80° - 60° = 20°$

Since angle $PQY = 20°$ and angle $QPY = 20°$, triangle PQY is isosceles.

Thus $PY = QY$

(sides opposite the equal angles in an isosceles triangle are equal)

19 5 minutes and 27 seconds past 1

20 OR $=$ OP (radii), so triangle ORP is isosceles

angle OPR $= \dfrac{180 - 20}{2} = 80°$ (angle sum of isosceles triangle)

OQ $=$ OP (radii), so triangle OQP is isosceles

angle OPR $= \dfrac{180 - 100}{2} = 40°$ (angle sum of isosceles triangle)

So angle OPQ $= \frac{1}{2} \times$ angle OPR and QP bisects angle OPR

Chapter 14 Answers

14.1 Get Ready

1 5 cm, 45 cm^2

Exercise 14A

1 a 40 cm^2 **b** 18 m^2 **c** 35 cm^2

   d 54 mm^2 **e** 30 cm^2 **f** 54 cm^2

2 Area 15 cm^2, area 25 cm^2, base 6 cm, area 32 cm^2, height 8 cm

Exercise 14B

1 a 56 cm^2 **b** 96 m^2 **c** 75 cm^2 **d** 98 cm^2

14.2 Get Ready

a $A = lw$

b $A = l^2$

c $A = \frac{1}{2}bh$

d $A = bh$

e $A = \frac{1}{2}(a + b)h$

Exercise 14C

1 a 28 m   **b** 37 m^2

2 60

3 30

4 6 cm

5 a i 32 m^2 **ii** 14 m^2

   b 24 m   **c** 13   **d** 9

6 a 66 cm   **b** 234 cm^2

7 30 cm^2

8 100 cm^2

14.3 Get Ready

1 a   **b**   **c**   **d**

Exercise 14D

1

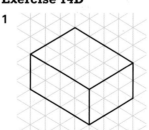

Answers

2 Any six out of:

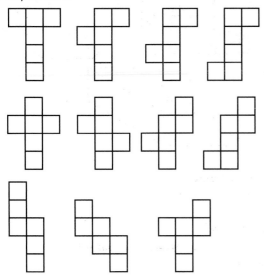

3 a Cylinder   **b** Cone
   c Triangular-based pyramid   **d** Square-based pyramid

4 a e.g.

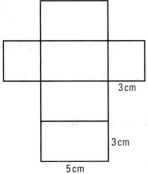

3 cm
3 cm
5 cm

b e.g.

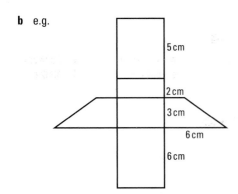

5 cm
2 cm
3 cm
6 cm
6 cm

14.4 Get Ready

1 a

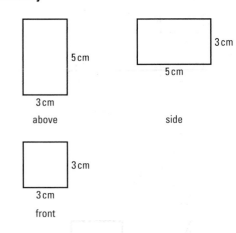

5 cm
3 cm
above

3 cm
5 cm
side

3 cm
3 cm
front

b

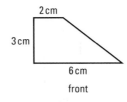

2 cm   4 cm
4 cm
above

3 cm
4 cm
side

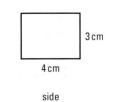

2 cm
3 cm
6 cm
front

Exercise 14E

1 a plan

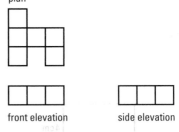

front elevation   side elevation

b plan

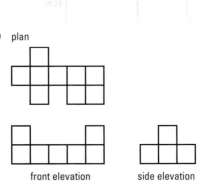

front elevation   side elevation

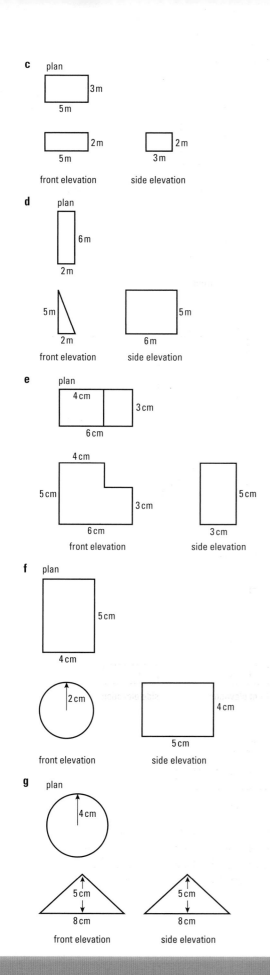

c
plan

5m — 3m

front elevation: 5m — 2m

side elevation: 3m — 2m

d
plan

2m — 6m

front elevation: 2m, 5m (triangle)

side elevation: 6m — 5m

e
plan

4cm, 6cm — 3cm

front elevation: 4cm, 5cm, 6cm, 3cm

side elevation: 3cm — 5cm

f
plan

4cm — 5cm

front elevation: circle, 2cm

side elevation: 5cm — 4cm

g
plan

circle, 4cm

front elevation: 5cm, 8cm (triangle)

side elevation: 5cm, 8cm (triangle)

2 a 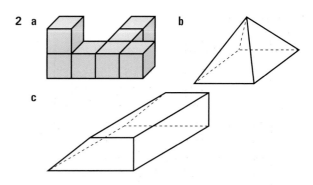 **b**

c

14.5 Get Ready

1 a $192\,\text{m}^3$ **b** $576\,\text{cm}^3$

Exercise 14F

1 a $396\,\text{cm}^3$ **b** $378\,\text{cm}^3$ **c** $204\,\text{cm}^3$
2 $400\,\text{cm}^3$

14.6 Get Ready

1 a a^3 **b** $2a^3$ **c** a^3

Exercise 14G

1 a $78\,\text{cm}^3$ **b** $2250\,\text{mm}^3$ **c** $0.498\,75\,\text{m}^3$
d $216\,\text{cm}^3$
2 a $225\,\text{cm}^3$ **b** $10\,500\,\text{cm}^3$ **c** $80.43\,\text{cm}^3$
d $84\,000\,\text{cm}^3$
3 $9\,\text{cm}$
4 Area $= \frac{1}{2}(x + 3x) \times 2x \times 2x = 8x^3\,\text{cm}^3$
5 $h = 4.5y$

14.7 Get Ready

1 $15\,\text{cm}^2$

Exercise 14H

1 a $10\,950\,\text{cm}^2$ **b** $11\,700\,\text{cm}^2$ **c** $3524\,\text{cm}^2$
d $3\,\text{m}^2$ **e** $684\,\text{cm}^2$ **f** $1392\,\text{cm}^2$
2 £94

14.8 Get Ready

1 (1.5, 3)

Answers

Exercise 14I

1 O (0, 0, 0), A (4, 0, 0), B (4, 0, 3), C (0, 0, 3), D (0, 2, 3), E (0, 2, 0), F (4, 2, 0), G (4, 2, 3)

2

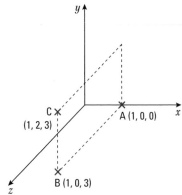

3 a O (0, 0, 0), A (4, 0, 0), B (4, 0, −3), C (0, 0, −3), D (0, −2, 0), E (4, −2, 0), F (4, −2, −3), G (0, −2, −3)
  b i (0, −2, −1.5)   **ii** (4, −1, −1.5)

4 (−2, −3, 0), (−2, −3, −1), (−2, 0, −1)

Review exercise

1

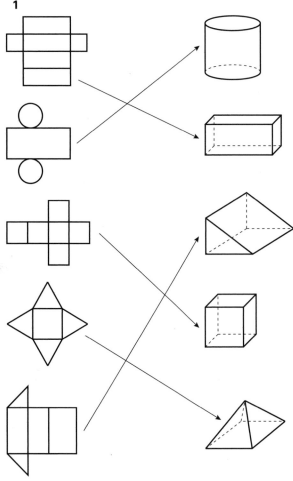

2 20 cm³

3

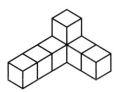

4 70 cm²

5 a

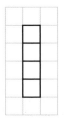

b

6 5 cm

7 150 boxes

8 a 5 cm   **b i** 19 boxes   **ii** 8 chocolates

9 a 118.12 cm²   **b** 83.52 cm³

10 96 cm²

11 110 cm²

12 a

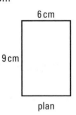

plan

front elevation   side elevation

  b 189 cm²

13 a 50 m of fencing
   b 270 m²

14 Skirting board: $3 \times 4 + 2 \times 2 + 1$, therefore cheapest
  $= 3 \times £30.50 + 1 \times £18.75 + 1 \times £14.00 = £124.25$
  Coving: 20 m, therefore cheapest
  $= 6 \times £27.50 + 1 \times £22.00 = £187.00$
  Total $= £311.25$

15 Area $= 28$ m², therefore $\frac{550}{28} = £19.64$ to spend per m²
  Amy can afford Natural Twist with either underlay or Medium Blend with Cushion.

16 Paving $= 20 \times £40 = £800$, grass $= 110 \times £15 = £1650$, total $= £2450$

17 Volume $= \frac{25}{2}(1 + 3) \times 10 = 500$ m³,
  so time $= 250$ minutes $= 4$ hours 10 minutes

18 a (5, 2, 0)   **b** (2.5, 1, 3)

19 a 315 cm³   **b** 0.6 g/cm³

Multiplication

1 £249.50
2 Yes, her nan will have to give her £35.80.
3 £48.46

Area

1 e.g.

2 £210
3 £248
4 e.g. 1 cm by 36 cm, perimeter 74 cm; 2 cm by 18 cm,
    perimeter 40 cm; 3 cm by 12 cm, perimeter 30 cm; 4 cm by
    9 cm, perimeter 26 cm; 6 cm by 6 cm, perimeter 24 cm
5 £8

Communication

1 Students' comparisons of costs for different numbers of
   months. For more than 18 months, 'pay as you go' is
   cheaper, otherwise monthly contract is cheaper.
2 215.04 seconds
3 13 cm by 19 cm (285 ppi by 285 ppi)
   15 cm by 23 cm (247 ppi by 235 ppi)
   20 cm by 30 cm (185 ppi by 180 ppi)
   Ranji should print the photo at 13 cm by 19 cm.

Energy efficiency

1 100 mm: Space Combi (Economy roll £200, Easy Roll £175,
   Space Combi £40)
   150 mm: Space Blanket (medium) (Economy roll £300,
   Space Blanket (medium) £168)
   200 mm: Space Combi (Economy roll £400, Space Blanket
   (thick) £210, Space Combi £80)
2 £15.18
3 100 weeks

Index

Index